"Fluid yet engaging, just like a good conversation over a pan of sizzling vegetables." —*New Republic*

◆◆

"[A] delightfully informative history of cooking and eating." —*ELLE Magazine*

◆◆

"A book to savour... You will never look at a kitchen knife in the same way again." —*Independent* (London)

◆◆

"Full of intriguing scholarship... Socially astute and funny." —*The New Yorker*

◆◆

"[An] ambitious, blenderized treatise." —*The Washington Post*

◆◆

"Witty, scholarly, utterly absorbing and fired by infectious curiosity." —*Observer* (London)

◆◆

"Wilson is a good tour guide.... [A] dizzying, entertaining ride." —*Wall Street Journal*

◆◆

"Mouthwatering history: broad in scope, rich in detail, stuffed with savory food for thought." —*Publishers Weekly* (starred review)

Praise for Bee Wilson's *Consider the Fork*

"Bee Wilson's supple, sometimes playful style in *Consider the Fork* . . . cleverly disguises her erudition in fields from archaeology and anthropology to food science. . . . Wilson's insouciant scholarship and companionable voice convince you she would be great fun to spend time with in the kitchen. . . . [She is a] congenial kitchen oracle."
—Dawn Drzal, *New York Times Book Review*

"Wilson . . . writes beautifully and has the academic chops to deliver what she promises. . . . Reading the book is like having a long dinner table discussion with a fascinating friend. At one moment, she's reflecting on the development of cast-iron cookware, then she's relating the history of the Le Creuset company and the public's changing tastes in color and then she's reminiscing about her mother-in-law's favorite blue pots."
—*Los Angeles Times*

"Wilson remains engaging, and nowhere as deeply or as smoothly as in *Consider the Fork*, where the information she has to juggle is at once gastronomic, cultural, economic, and scientific. . . . Everything in Bee Wilson's pithy book brings you back to the kitchen: her histories of weights and measures and pots and pans; her observations on the domestication of fire and ice; . . . her homey riffs on small, exasperating 'technologies' like egg timers, cake molds, tongs, and toasters."—*The New Yorker*

"Substantial and entertaining. . . . Wilson belongs to a rare breed: the academic who can write. This book is dense with research, all of it rendered highly palatable. . . . The history comes in delicious nuggets of the kind that one immediately wants to pass around in conversation."
—*Mail on Sunday* (London)

"The path from Stone Age flints to sous-vide machines whirs so smoothly that I found myself re-reading passages just to trace how the author managed to work in a Victorian copper batterie de cuisine along the way."
—*The Washington Post*

"Each chapter is a discursive journey from past to present, across continents and cultures, opening our eyes to the wealth of material in our own kitchens. . . . [A] wide-ranging book, where . . . scholarship is always worn lightly. . . . Wilson is therefore equally at ease writing about recipes or economics, manners or morals."—*New York Review of Books*

"A delightful compendium of the tools, techniques and cultures of cooking and eating. Be it a tong or a chopstick, a runcible spoon or a cleaver, Bee Wilson approaches it with loving curiosity and thoroughness. . . . But as well as providing wry insights into the psychology of cooks down the ages, *Consider the Fork* is infused with a sense that every omelet, cup of coffee, meringue or tea cake is steeped in tradition and ancient knowledge, and that that is partly what makes cooking one of life's joys."—*Spectator* (London)

"[A] wide-ranging historical road map of the influence of culture on cuisine. . . . It is easy and delightful to get swept up in Wilson's zeal. And the rejection of a traditional narrative arc does not indicate a lack of structure; rather, the book's horizontal shape is a choice that suits its material. . . . Cooking is full of paradoxes. It is art and science, ancient and modern, fundamental and trivial, easy and difficult. Wilson presents these dissonances in their entirety, making no show of resolving them. In the end, her tone suggests that she writes about food for the same reason we read about it: sheer pleasure and lighthearted fascination. The big questions are just seasoning for the soup."—*New Republic Online*

"Wilson celebrates the unsung implements that have helped shape our diets through the centuries. After devouring this delightful mix of culinary science and history, you'll never take a whisk for granted again."—*Parade*

"Wilson . . . skillfully turns a potentially dull subject into one of wit and wisdom. Nor does she lose touch with the human element that has drawn so many into the world of cooking and the universal subject of food. After all, a knife is only as good as the cook who wields it. Wilson packs *Consider the Fork* with as many bits of cultural history trivia as an overstuffed utensil drawer."—*Christian Science Monitor*

"What new intellectual vistas remain to be conquered by the food obsessive? . . . The erudite and witty food writer Bee Wilson has spotted a gap in the market. . . . [Her] argument is clear and persuasive."
—*Guardian* (London)

"Endlessly fascinating."—*New Statesman* (London)

"A book to keep at your side as you cook. . . . Wilson serves up brisk histories of everything you use in the kitchen."—*The Daily Beast*

"You know that corner cupboard full of kitchen gadgets that promised to transform our lives? . . . In that corner cupboard lies a rich history of technological promise and dashed hopes. One that, until Bee Wilson's recent release, *Consider the Fork*, nobody has thought to tell. . . . Clearly-written and methodically researched, *Consider the Fork* fills a real void in culinary literature."—*Toronto Star*

"Wilson's tour of the kitchen explores all the essential elements of domestic cookery through the ages. . . . Wilson's book is diligently researched and she has a sharp eye for a vivid historical detail."—*Daily Mail* (London)

"Like all the best books on apparently simple everyday commodities, this is of course really a gripping story of millennia of human ingenuity. . . . *Consider the Fork* wears its impressive research lightly."—*Observer* (London)

"Wilson shifts the focus from the foods people ate to the technology behind their preparation, tracing how humble kitchen implements such as forks, whisks, pots, and stoves shaped our diets, our societies, and our bodies. In Wilson's hands, even hot water becomes interesting."
—*Discover Magazine*

"Open[s] windows on the dynamic interplay of science, technology and the culinary arts in history. . . . *Consider the Fork* delves into the chewy past of kitchen technology."—*Nature*

"Focusing on culinary tools, the author hopscotches through human history around the globe, recording both strides and stumbles."
—*Washington Times*

"Wilson's spirited history of kitchen implements ranges from the humble wooden spoon to the cutting-edge sous vide machine. A British food writer and historian, Wilson is learned and personal, wise and charming. . . . There are complex investigations at work in Wilson's book; it's nominally about things in our cabinets and on our shelves, but it's really about family, labor, technology, sensation. . . . From such ingredients an enchanting book is made."—*Smithsonian Magazine*

"One of the delights of *Consider the Fork* is that [Wilson's] fascination with the history of food is balanced by the pleasure she takes in preparing dishes herself, watching others do so and, best of all, tasting the results. Ms. Wilson's design critiques of different utensils, from the humble wooden spoon to a snazzy sous-vide water bath, are all the more convincing for being made by a knowledgeable and passionate cook, who isn't afraid to admit to her failures, yet longs for delicious successes."
—Alice Rawsthorn, NewYorkTimes.com

"Some of humanity's least sung but most vital gadgets are celebrated in this delicious history of cooking technology. . . . Wilson is erudite and whip-smart, but she always grounds her exploration of technological change in the perspective of the eternal harried cook—she's been one—struggling to put a meal on the table."—*Publishers Weekly*, Starred Review

"At every turn, Wilson's history of the technology of cooking and eating up-ends another unexamined tradition, revealing that utensils and practices now taken for granted in kitchen and at table have long and remarkable histories. . . . Wilson's book teems with . . . delightful insights."
—*Booklist*, Starred Review

"At the risk of trotting out a cliché, Brit writer Wilson's book truly is food for thought. (And fun to read, too)."—*New York Post*

"If you are open to being entertained and instructed by the history of food, then Bee Wilson couldn't be happier to oblige. In *Consider the Fork*, she explores the ways in which kitchen tools and techniques affect what and how we eat, with the same owlish brio and dry humor that Jane Grigson brought to vegetables and charcuterie. . . . [A] smart, regaling survey."
—*Barnes & Noble Review*

"Like a well-planned meal, *Consider the Fork* provides a variety of fare that will entertain and educate foodies of any variety. . . . The result of [Wilson's] combination of sophisticated humor and scholarship is an enjoyable tale about the very essence of existence and civilization."—*Roanoke Times*

"Chockfull of revelations that any cooking enthusiast will eat up with a spoon."—*New York Journal of Books*

"Wilson is the ideal guide. Her scholarship is substantive, nerdy, detailed enough. . . . That's precisely the pleasure of reading her."
—*New City* (Chicago)

"Wilson's sprightly, knowledgeable voice skips nimbly through the narratives of pots and pans, knives, grinding implements and eating utensils, working up to the theme of the kitchen as a whole. . . . Don't be surprised if you find yourself sitting up at night with *Consider the Fork*, unable to turn out the light until you find out how storing and shipping ice became viable. You will never again walk into your kitchen without thinking of the rich history represented by even the humble fork."—*Shelf Awareness*

"[Wilson]'s at her sparkling best when unearthing curious histories about the role these inventions played in the evolution of man. She serves up her impressive research in easy-to-digest nuggets, making the chronicle of even the dullest kitchen aid a palatable treat."—*Metro UK*

"In the lively prose of a seasoned journalist, Wilson blends personal reminiscences with well-researched history to illustrate how the changing nature of our equipment affects what we eat and how we cook. . . . Rarely has a book with so much information been such an entertaining read."
—*Kirkus Reviews*

"This scholarly and witty book, packed full of fascinating information and thrilling insights, is as enlightening as it is a joy to read."—Claudia Roden

"Mind meets kitchen: Bee Wilson sizes up every kitchen implement from the wooden spoon to the ergonomic Microplane, and gives us its history, including versions that led up to each object but did not survive for lack of fitness. Her climax is the kitchen, the room itself, the affluent modern version of which has never been 'so highly designed; so well equipped; so stylish; or so empty.' She conducts us on a sobering, entertaining, and instructive tour."—Margaret Visser, author of *Much Depends on Dinner*

"*Consider the Fork* is a terrific delve into the history and modern use of kitchen tools so familiar that we take them for granted and never give them a thought. Bee Wilson places kitchen gadgets in their rich cultural context. I, for one, will never think about spoons, measuring cups, eggbeaters, or chopsticks in the same way again."
—Marion Nestle, Professor of Nutrition, Food Studies, and Public Health, New York University, and author of *What to Eat*

CONSIDER
the FORK

Also by Bee Wilson

Sandwich
Swindled
The Hive

CONSIDER
the FORK

A History of
How We Cook and Eat

BEE WILSON

with illustrations by
ANNABEL LEE

BASIC BOOKS
A MEMBER OF THE PERSEUS BOOKS GROUP
NEW YORK

Books published by Basic Books are available at special discounts
for bulk purchases in the United States by corporations, institutions, and
other organizations. For more information, please contact the Special Markets
Department at the Perseus Books Group, 2300 Chestnut Street,
Suite 200, Philadelphia, PA 19103, or call (800) 810-4145,
ext. 5000, or e-mail special.markets@perseusbooks.com.

Designed by Linda Mark
Text set in Fairfield Light by the Perseus Books Group

The Library of Congress has cataloged the hardcover edition as follows:
Wilson, Bee.
Consider the fork : a history of how we cook and eat /
Bee Wilson ; with illustrations by Annabel Lee.
p. cm.
Includes bibliographical references and index.
ISBN 978-0-465-02176-5 (hardback)—ISBN 978-0-465-03332-4 (e-book)
1. Kitchen utensils—History. 2. Cooking—History.
3. Dinners and dining—History. I. Title.
TX656.W56 2012
643'.3—dc23
2012016283

ISBN 978-0-465-05697-2 (paperback)

10 9 8 7 6 5 4 3 2 1

For my Mother

CONTENTS

INTRODUCTION

A WOODEN SPOON—MOST TRUSTY AND LOVABLE OF KITCHEN implements—looks like the opposite of "technology," as the word is normally understood. It does not switch on and off or make funny noises. It has no patent or guarantee. There is nothing futuristic or shiny or clever about it.

But look closer at one of your wooden spoons (I'm assuming you have at least one, because I've never been in any kitchen that didn't). Feel the grain. Is it a workmanlike beech factory spoon or a denser maple wood or olive wood whittled by an artisan? Now look at the shape. Is it oval or round? Slotted or solid? Cupped or flat? Perhaps it has a pointy part on one side to get at the lumpy bits in the corner of the pan. Maybe the handle is extrashort, for a child to use, or extralong, to give your hand a position of greater safety from a hot skillet. Countless decisions—economic and social as well as those pertaining to design and applied engineering—will have gone into the making of this object. And these in turn will affect the way this device enables you to cook. The wooden spoon is a quiet ensemble player in so many meals that we take it for granted. We do not give it credit for the eggs it has scrambled, the chocolate it has helped to melt, the onions it has saved from burning with a quick twirl.

The wooden spoon does not look particularly sophisticated—traditionally, it was given as a booby prize to the loser of a competition—but it has science on its side. Wood is nonabrasive and therefore gentle on pans—you can scrape away without fear of scarring the metal surface. It is nonreactive: you need not worry that it will leave a metallic taste or that its surface will degrade on contact with acidic citrus or tomatoes. It is also a poor conductor of heat, which is why you can stir hot soup with a wooden spoon without burning your hand. Above and beyond its functionality, however, we cook with wooden spoons because we always have. They are part of our civilization. Tools are first adopted because they meet a certain need or solve a particular problem, but over time the utensils we feel happy using are mainly determined by culture. In the age of stainless steel pans, it is perfectly possible to use a metal spoon for stirring without ruining your vessels, but to do so feels obscurely wrong. The hard metal angles smash your carefully diced vegetables and the handle does not grip so companionably as you stir. It clanks disagreeably, in contrast to the gentle tapping of wood.

In this plastic age, you might expect that we would have taken to stirring with synthetic spatulas, especially because wooden spoons don't do well in dishwashers (over many washes, they tend to soften and split); but on the whole, this is not so. I saw a bizarre product in a kitchenware shop recently: "wooden silicone spoons," on sale for eight times the price of a basic beech spoon. They were garishly colored, heavy plastic kitchen spoons in the shape of a wooden spoon. Apart from that, there was nothing wooden about them. Yet the manufacturers felt that they needed to allude to wood to win a place in our hearts and kitchens. There are so many things we take for granted when we cook: we stir with wooden spoons but eat with metal ones (we used to eat with wood, too); we have strong views on things that should be served hot and things that must remain raw. Certain ingredients we boil; others, we freeze or fry or grind. Many of these actions we perform instinctively, or by obediently following a recipe. It is perfectly clear to anyone who prepares Italian food that a risotto should be cooked with the gradual addition of liquid, whereas pasta needs to be boiled fast in an excess of water, but why?* Most aspects of cooking are far less obvious than they first appear; and there is almost always another way of doing things. Think of the utensils that were not adopted, for whatever reason: the water-powered egg whisk, the magnet-operated spit roaster. It took countless inventions, small and large, to get to the well-equipped kitchens we have now, where our old low-tech friend the wooden spoon is joined by mixers, freezers, and microwaves; but the history is largely unseen and unsung.

Traditional histories of technology do not pay much attention to food. They tend to focus on hefty industrial and military developments: wheels and ships, gunpowder and telegraphs, airships and radio. When food is mentioned, it is usually in the context of agriculture—systems of tillage and irrigation—rather than the domestic work of the kitchen. But there is just as much invention in a nutcracker as in a bullet. Often, inventors have been working on something for military use, only to find that its best use is in the kitchen. Harry Brearley was a Sheffield man who invented stainless steel in 1913 as a way of improving gun barrels; inadvertently, he improved the world's cutlery. Percy Spencer, creator of the microwave oven, was working on naval radar systems when he happened upon an entirely new method of cooking. Our kitchens owe much to the brilliance of science, and a cook experimenting with mixtures at the stove is often not very different from a chemist in the lab: we add vinegar to red cabbage to fix the color and use baking soda to counteract the acidity of lemon in a cake. It is wrong to suppose, however, that technology is

---

* You might reply: because risotto needs to be starchy and creamy, whereas slippery pasta benefits from having some of its starch washed away in the water. But this still begs the question. Pasta can be delicious cooked risotto-style, particularly the small rice-shaped orzo, with the incremental addition of wine and stock. Equally, risotto-style rice can be very good with a single large addition of liquid at the beginning, as with paella.

just the appliance of scientific thought. It is something more basic and older than this. Not every culture has had formal science—a form of organized knowledge about the universe that starts with Aristotle in the fourth century BC. The modern scientific method, in which experiments form part of a structured system of hypothesis, experimentation, and analysis is as recent as the seventeenth century; the problem-solving technology of cooking goes back thousands of years. Since the earliest Stone Age humans hacking away at raw food with sharpened flints, we have always used invention to devise better ways to feed ourselves.

The word *technology* comes from the Greek. *Techne* means an art, skill, or craft, and *logia* means the study of something. Technology is not a form of robotics but something very human: the creation of tools and techniques that answer certain uses in our lives. Sometimes technology can mean the tools themselves; other times it refers to the inventive know-how that made the tools possible, or the fact that people use these particular tools and not others. Scientific discovery does not depend on usage for its validity; technology does. When equipment falls out of use, it expires. However shrewdly designed it may be, an eggbeater does not fully achieve its purpose until someone picks it up and beats eggs.

Consider the Fork is an exploration of the way the implements we use in the kitchen affect what we eat, how we eat, and what we feel about what we eat. Food is the great human universal. Nothing is certain in this world except death and taxes, the saying goes. It should really be death and food. Plenty of people avoid taxes (not earning any money is one way, but certainly not the only one). Some live without sex, that other fact of life. But there is no getting beyond food, which is a fuel, a habit, a higher pleasure, and a base need, the thing that gives pattern to our days or that gnaws us with its lack. Anorexics may try to escape it, but for as long as you live, hunger is inescapable. We all eat. Yet the ways in which we have satisfied this basic human need have varied dramatically at different times and places. The things that make the biggest difference are the tools we use.

Most days, my breakfast consists of coffee; toast, butter, marmalade; and orange juice, if the children haven't drunk it all. Described like this, as bare ingredients, it is a meal that could belong to any moment of the past three hundred and fifty years. Coffee has been consumed in England since the mid-seventeenth century; oranges for the juice and the marmalade since 1290. Toasted bread and butter are both ancient. The devil, however, is in the details.

To make the coffee, I do not boil it for twenty minutes and then clarify it with isinglass (fish bladder), as I might have done in 1810; I do not make it in a "scientific Rumford percolator," as some did in 1850; I do not make it in a jug with a wooden spoon, pouring cold water over the hot grounds to make them fall to the bottom in the Edwardian style; I do not make it in an electric coffeemaker, as I might still if I lived in the States; I do not pour hot water over an acrid spoonful of instant as in student years; and I do not generally make it in a French press cafetière, though I did in the 1990s. I am an early twenty-first-century coffee obsessive (but not obsessive enough, yet, to have invested in a state-of-the-art Japanese siphon brewer). I grind my beans (fair trade) superfine in a burr grinder and make myself a "flat white" (an espresso, steamed milk poured over the top), using an espresso machine and a range of utensils (coffee scoop, tamper, steel milk pitcher). On good mornings, after ten minutes or so of concentrated effort, the technology works, and the coffee and milk meld into a delicious foamy drink. On bad mornings, they explode all over the floor.

Toast, butter, and marmalade were known and loved by the Elizabethans. But Shakespeare never ate toast such as mine, cut from a whole-grain loaf baked in an automatic bread maker, toasted in a four-slot electric toaster, and eaten off a white dishwasher-safe china plate. Nor did he know the joys of spreadable butter and high-fruit marmalade, both of which indicate the presence in my household of a large and fully functioning refrigerator. Besides, Shakespeare's marmalade would probably have been made with quinces, not oranges. My butter is not rancid or too hard—as I remember almost all butter being when I was a child in the 1970s and 1980s. I spread it with a stainless steel knife, which leaves no metallic tang and does not react with the fruit sugars in the marmalade.

As for the orange juice, the technology behind it seems the simplest of all—take oranges, squeeze juice—but is probably the most complicated. Unlike the Edwardian housewife, who laboriously squeezed oranges in a conical glass squeezer, I usually pour my juice from a Tetra Pak carton (first launched as Tetra Brik in 1963). Although the ingredients list only oranges, the juice will have been made using a bewildering array of industrial techniques, the fruit crushed with hidden enzymes and strained with hidden clarifiers and pasteurized and chilled and transported from country to country, all for my breakfast pleasure. The fact that the juice does not pucker my mouth with bitterness is thanks to a female inventor, Linda C. Brewster, who in the 1970s was granted four patents for "debittering" orange juice by reducing the presence of acrid limonin.

This particular meal could only have been consumed in this particular way for a very short moment in history. The foods we eat speak of the time and the place we inhabit. But to an even greater extent, so do the tools we use to make and consume them. We are often told that we live in a "technological age." This is usually a way of saying: we have a lot of computers. But 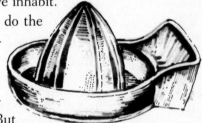 every age has its technology. It does not have to be futuristic. It can be a fork, a pot, or a simple measuring cup.

Sometimes, kitchen tools are simply a way of enhancing the pleasure of eating. But they can also be a matter of basic survival. Before the adoption of cooking pots, around 10,000 years ago, the evidence from skeletons suggests that no one survived into adulthood having lost all their teeth. Chewing was a necessary skill. If you couldn't chew, you would starve. Pottery enabled our ancestors to make food of a drinkable consistency: porridgy, soupy concoctions, which could be eaten without chewing. For the first time, we start to see adult skeletons without a single tooth. The cooking pot saved these people.

The most versatile technologies are often the most basic. Some, like the mortar and pestle, endure for tens of thousands of years. The pestle began as an ancient tool for processing grain but successfully adapted itself to grinding everything from *pistou* in France to curry paste in Thailand. Other devices have proved less flexible, for instance, the 1970s chicken brick, enjoying a brief vogue only to end up on the junk heap when people tired of the food in question. Some tools, such as spoons and microwaves, are used the world over. Others are very specific to a place, for example, the *dolsot*, a sizzling hot stone pot in which Koreans serve one particular dish: *bibimbap*, a mixture of sticky rice, finely sliced vegetables, and raw or fried egg; the bottom layer of rice becomes crispy with the heat of the *dolsot*.

This book is about high-tech gadgets, but it is also about the tools and techniques we don't tend to think about so much. The technology of food matters even when we barely notice it is there. From fire onward, there is a technology behind everything we eat, whether we recognize it or not. Behind every loaf of bread, there is an oven. Behind a bowl of soup, there is a pan and a wooden spoon (unless it comes from a can, another technology altogether). Behind every restaurant-kitchen foam, there will be a whipping canister, charged with N_2O. Ferran Adrià's El Bulli in Spain, which, until it closed in 2011, was the most celebrated restaurant in the world, could not have produced its menu without sous-vide machines and centrifuges, dehydrators, and Pacojets. Many people find these novel tools alarming. As new kitchen technologies have emerged, there have always been voices suggesting that the old ways were best.

Cooks are conservative beings, masters of quiet repetitive actions that change little from day to day or year to year. Entire cultures are built around cooking food one way and not another. A true Chinese meal, for example, cannot be cooked without the *tou*, the cleaver-shaped knife that reduces ingredients to small, even morsels, and the wok, for stir-frying. Which comes first, the stir-fry or the wok? Neither. To get at the logic of Chinese cuisine, we have to go even further back and consider cooking fuel: a quickly made wok-cooked meal was originally the product of firewood scarcity. Over time, however, equipment and food become so bound together you can't say when one starts and the other ends.

It is only natural that cooks should perceive kitchen innovation as a personal attack. The complaint is always the same: you are destroying the food we know and love with your newfangled ways. When commercial refrigeration became a possibility in the late nineteenth century, it offered great advantages, both to consumers and industry. Fridges were especially useful for selling perishable substances such as milk, which had previously been the cause of thousands of deaths every year in the big cities of the world. Refrigeration benefited traders, too, creating a longer window in which they could sell their food. Yet there was a widespread terror of this new technology, from both sellers and buyers. Consumers were suspicious of food that had been kept in cold storage. Market traders, too, did not know what to make of this new chill. In the 1890s at Les Halles, the huge central food market in Paris, the sellers felt that refrigeration would spoil their produce. And at some level, they were right, as anyone who has ever compared a tomato at room temperature with one from the fridge can confirm: the one (assuming it's a good tomato) is sweetly fragrant and juicy; the other is woolly, metallic, and dull. Every new technology represents a trade-off: something is gained, but something is also lost.

Often, the thing lost is knowledge. You don't need such good knife skills once you have a food processor. Gas and electric ovens and the microwave mean you need no knowledge of how to get a fire going and keep it ablaze. Until around a hundred years ago, management of a fire was one of the dominant human activities. That has gone (and a good thing, too, if you think of all the tedious hours of the day it consumed, all the other activities it precluded). The larger question is whether the existence of cooking technologies that entail only minimal human input has led to the death of culinary skills. In 2011, a survey of 2,000 British young people from age eighteen to twenty-five found that more than half said that they had left home without the ability to cook even a simple recipe such as Spaghetti Bolognese. Microwaves plus convenience foods offer the freedom of being able to feed yourself with a few pushes of a button. But it's not such a great advance if you lose all concept of what it would mean to make a meal for yourself.

Sometimes, though, it takes a new technology to make us appreciate an old one. The knowledge that I can make hollandaise in thirty seconds in the blender enhances the pleasure of doing it the old way, with a double boiler and a wooden spoon, the butter added to the yolks piece by tiny piece.

The equipment of the kitchen can seem unimportant compared to the history of food itself. It is all very well fussing over the niceties of table settings and jelly molds, but what does this matter compared to a basic hunger for bread? Perhaps this explains why kitchen tools have been so neglected in histories of food. Culinary history has become a hot subject over the past two decades. But the focus of these new histories, with a few notable exceptions, has overwhelmingly been ingredients rather than technique: *what* we cooked rather than *how* we cooked it. There have been books on potatoes, cod, and chocolate, and histories of cookbooks, restaurants, and cooks. The kitchen and its tools are more or less absent. As a result, half the story is missing. This matters. We change the texture, the taste, the nutritional content, and the cultural associations of ingredients simply by using different tools and techniques to prepare them.

Beyond this, we human beings have been changed by kitchen technology—the *how* of food as well as the *what*. I don't just mean this in a "my dream kitchen changed my life" kind of way, though it is true that changes in kitchen tools have gone hand in hand with vast social changes. Take the relationship between labor-saving devices and servants. The story here is one of technological stagnation. There was very little interest in eliminating the grind of cooking for the many centuries when well-off kitchens came with an abundance of human labor to take the strain. Electric food processors and blenders are genuinely liberating tools. Arms no longer have to ache to produce kibbe in Lebanon or ginger-garlic puree in India. So many meals that were once seasoned with pain are now trouble free.

Kitchen tools have changed us in more physical ways. There is good evidence to suggest that the current obesity crisis is caused, in part, not by what we eat (though this is of course vital, too) but by the degree to which our food has been processed before we eat it. It is sometimes referred to as the "calorie delusion." In 2003, scientists at Kyushu University in Japan fed one group of rats hard food pellets and another group softer pellets. In every other respect the pellets were identical: same nutrients, same calories. After twenty-two weeks, the rats on the soft-food diet had become obese, showing that texture is an important factor in weight gain. Further studies involving pythons (eating ground cooked steak, versus intact raw steak) confirmed these findings. When we eat chewier, less processed foods, it takes us more energy to digest them, so the number of calories our body receives is less. You will get more energy from a slow-cooked apple puree than a crunchy raw apple, even if the calories on paper are identical. Food labels, which still display nutritional information in crude terms of calories (according to the Atwater convention on nutrition developed in the late nineteenth century), have not yet caught up with this, but it is a stark example of how the technology of cooking really matters.

In many ways, the history of food *is* the history of technology. There is no cooking without fire. The discovery of how to harness fire and the consequent art of cooking was what enabled us to evolve from apes to *Homo erectus*. Early hunter-gatherers may not have had KitchenAids and "Lean, Mean Grilling Machines," but they still had their own version of kitchen technology. They had stones to pound with and sharpened stones to cut with. With dexterous hands, they would have known how to gather edible nuts and berries without getting poisoned or stung. They hunted for honey in lofty rock crevices and used mussel shells to catch the dripping fat from a roasting seal. Whatever else was lacking, it was not ingenuity.

This book tells the story of how we have tamed fire and ice, how we have wielded whisks, spoons, graters, mashers, mortars and pestles, how we have used our hands and our teeth, all in the name of putting food in our mouths. There is hidden intelligence in our kitchens, and the intelligence affects how we cook and eat. This is not a book about the technology of agriculture (there are other books about that). Nor is it very much about the technology of restaurant cooking, which has its own imperatives. It is about the everyday sustenance of domestic households: the benefits that different tools have brought to our cooking and the risks.

We easily forget that technology in the kitchen has remained a matter of life and death. The two basic mechanisms of cooking—slicing and heating—are fraught with danger. For most of human history, cooking has been a largely grim business, a form of dicing with danger in a sweaty, smoky, confined space. And it still is in much of the world. Smoke, chiefly from indoor cooking fires, kills 1.5 million people every year in the developing world, according to the World Health Organization. Open hearths were a major cause of death in Europe, too, for centuries. Women were particularly at risk, on account of the terrible combination of billowing skirts, trailing sleeves, and open fires with bubbling cauldrons hung over them. Professional chefs in rich households until the seventeenth century were almost universally men, and they often worked naked or just in undergarments on account of the scorching heat. Women were confined to the dairy and scullery, where their skirts didn't pose such a problem.

One of the greatest revolutions to take place in the British kitchen came with the adoption of enclosed brick chimneys and cast-iron fire grates, over the course of the sixteenth and seventeenth centuries. A whole new set of kitchen implements emerged, in tandem with this new control of the heat source: suddenly, the kitchen was not such a foul and greasy place to be, and gleaming brass and pewter pots took over from the blackened old cast iron. The social consequences were huge, too. At last, women could cook food without setting fire to themselves. It is no coincidence that a generation or so after enclosed oven ranges became the norm, the first cookbooks written by women for women were published in Britain.

Kitchen tools do not emerge in isolation, but in clusters. One implement is invented and then further implements are needed to service the first one. The birth of the microwave gives rise to microwave-proof dishes and microwavable plastic wrap. Freezers create a sudden need for ice cube trays. Nonstick frying pans necessitate nonscratch spatulas. The old open-hearth cookery went along with a host of related technologies: andirons or brand-irons to stop logs from rolling forward; gridirons for toasting bread; hasteners— large metal hoods placed in front of the fire to speed up cooking; various spit-jacks for turning roasting meat; and extremely long-handled iron ladles, skimmers, and forks. With the end of open-hearth cookery, all of these associated tools vanished, too.

For every kitchen technology that has endured—like the mortar and pestle—there are countless others that have vanished. We no longer feel the need of cider owls and dangle spits, flesh-forks and galley pots, trammels, and muffineers, though in their day, these would have seemed no more superfluous than our oil drizzlers, electric herb choppers, and ice-cream scoops. Kitchen gizmos offer a fascinating glimpse into the preoccupations of any given society. The Georgians loved roasted bone marrow and devised a special silver spoon for eating it. The Mayans lavished great artistry on the gourds from which chocolate was drunk. If you walk around our own kitchenware shops, you would think that the things we are really obsessed with in the West right now are espresso, panini, and cupcakes.

Technology is the art of the possible. It is driven by human desire— whether the desire to make a better cupcake or the simple desire to stay alive—but also by the materials and knowledge available at any given time. Food in cans was invented long before it could easily be used. A patent for Nicolas Appert's revolutionary new canning process was issued in 1812, and the first canning factory opened in Bermondsey, London, in 1813. Yet it would be a further fifty years before anyone managed to devise a can opener.

The birth of a new gadget often gives rise to zealous overuse, until the novelty wears off. Abraham Maslow, a guru of modern management, once said that to the man who has only a hammer, the whole world looks like a nail. The same thing happens in the kitchen. To the woman who has just acquired an electric blender, the whole world looks like soup.

Not every kitchen invention has been an obvious improvement on what came before. My kitchen cupboards are graveyards of passions that died: the electric juicer I thought would change my life until I discovered I couldn't bear to clean it; the rice cooker that worked perfectly for a year and then suddenly burned every batch it made; the Bunsen burner with which, I imagined, I would create a series of swanky crème brûlées for dinner parties I never actually gave. We can all think of examples of more or less pointless pieces of culinary equipment—the melon baller, the avocado slicer, the garlic peeler—to which we can only respond: what was wrong with a spoon, a knife, or fingers? Our cooking benefits from much uncredited engineering, but there have also been gadgets that create more problems than they solve, and others that work perfectly well, but at a human cost.

Historians of technology often quote Kranzberg's First Law (formulated by Melvin Kranzberg in a seminal essay in 1986): "Technology is neither good nor bad; nor is it neutral." This is certainly true in the kitchen. Tools are not neutral objects. They change with changing social context. A mortar and pestle was a different thing for the Roman slave forced to pound up highly amalgamated mixtures for hours on end for his master's enjoyment than it is for me: a pleasing object with which I make pesto for fun, on a whim.

At any given time, we do not necessarily get the tools that would—in absolute terms—make our food better and our lives easier. We get those that we can afford and those that our society can accept. From the 1960s onward, a series of historians pointed out the irony that the amount of time American women spent on housework, including cooking, had remained constant since the mid-1920s, despite all of the technological improvements that came on the market over those four decades. For all the dishwashers, electric mixers, and automatic garbage disposals, women were working as hard as ever. Why? Ruth Schwartz Cowan, in her campaigning history *More Work for Mother*, noted that in pure technical terms, there was no reason America should not have communal kitchen arrangements, sharing out the cooking among several households. But this technology was never widely explored because the idea of public kitchens is socially unacceptable: most of us generally like to live in smaller family units, however irrational that may be.

Kitchen gadgets—especially the fancy, expensive kind that are sold through the shopping channels—advertise themselves with the promise that they will change your life. Often, however, your life is changed in ways that you did not expect. You buy an electric mixer, which makes it incredibly quick and easy to make cakes. And so you feel that you *ought* to make cakes, whereas before you acquired the mixer, making cakes was so laborious that you were happy to buy them. In fact, therefore, the mixer has cost you time, rather than saving it. There's also the side effect that in making room for the mixer, you have lost another few precious inches of counter space. Not to mention the hours you will spend washing the bowl and attachments and mopping the flour that splatters everywhere as it mixes.

Just because a technology is there doesn't mean we have to use it. There is almost no kitchen tool so basic that someone somewhere hasn't rejected it as "not worth the trouble." Yet it is true that most of our kitchens contain far more stuff than we need. When you reach the point where you can't open the utensils drawer because it is so jammed with rolling pins, graters, and fish slicers, it's time to shed a few technologies. In extremis, a skilled cook could manage pretty well with nothing but a sharp knife, a wooden board, a skillet, a spoon, and some kind of heat source.

But would you want to? Part of what makes cooking exciting is how this eternal business of putting food in our mouths subtly alters from decade to decade. Ten or twenty years from now, I'm sure my breakfast will have changed, even if I cling to the same coffee, toast, butter, marmalade, and juice. If the past is anything to go by, some of the techniques that once seemed so right will suddenly seem out of kilter. I am already starting to regret the bread maker—such an ugly object, and there's always a hole in the middle of the loaf from the paddle—and returning to the low-tech business of buying good sourdough from a baker or making my own by hand. My espresso machine finally broke while I was writing this book, and I've just discovered the AeroPress, an amazing, inexpensive manual device that makes inky-dark coffee essence using air pressure. With the marmalade, I am tempted to go electric and get an automatic jam maker.

As for the rest, who can say if comfortable breakfasts like mine will exist a few years from now? Oranges from Florida may become unaffordable as wind farms replace citrus farms to meet rising energy needs. Butter may go the same way (I pray this never happens) as dairy land is diverted to more efficient use growing plant foods. Or perhaps in the techno-kitchen of the future, we will all be breakfasting off "baconated grapefruit" and "caffeinated bacon," as Matt Groening imagines in an episode of *Futurama*.

One thing is sure. We will never get beyond the technology of cooking itself. Sporks may come and go, microwaves rise and fall. But the human race will always have kitchen tools. Fire, hands, knives; we will always have these.

POTS AND PANS

Cook, little pot, cook.
>THE BROTHERS GRIMM, "Sweet Porridge"

Boiled food is life, roast food death.
>CLAUDE LÉVI-STRAUSS, *The Origin of Table Manners*

THE COOKING POT I USE MOST OFTEN IS NOTHING amazing. I got it mail order on special offer as part of a ten-piece set from a Sunday supplement in the early days of married life, when owning our own set of gleaming pans, all matching—as opposed to the assorted chipped-enamel vessels of student days—seemed mysteriously grown-up. The set was stainless steel. "Order now and save $$$$ plus receive a free milk pan!" said the ad. So I did. They have seen us good, these pans. We even used the free milk pan for a long time, to warm milk for my daughter's morning cereal, though annoyingly, it lacked a pouring spout, so a bit of the milk sometimes sloshed out onto the work surface. And then, one fine morning, the handle fell off. Still, they are trusty pans, on the whole. Thirteen years later, I haven't managed to

destroy a single one outright. They have withstood burnt risotto, neglected stews, sticky caramels. Stainless steel may not conduct heat
as well as copper; it may not retain warmth as well as cast iron or
clay; it may not be as beautiful as enameled iron; but it comes into
its own at the dishwashing stage.

In particular, we have gotten stalwart service from a medium-
sized lidded pan with two small looped handles. The technical term
for it, I believe, is a sauce pot,
though a better word for it
would be the French *fait-tout*
because it really does do everything. It gets
dragged onto the cooktop for morning
porridge and again for evening rice. It has
known the creamy blandness of custards and rice puddings; the
spicy heat of curry; and numberless soups, from smooth green watercress to peppery minestrone. It is my workaday pan. Too small for
pasta or stock making, it does the boiling jobs I don't think twice
about. Switch on electric kettle; pour heated water into saucepan;
add salt; throw in broccoli florets/green beans/cobs of corn; lid on or
off depending on my mood; boil for a few minutes; drain in a colander; job done. There is nothing challenging or groundbreaking about
this process. The French generally deride such cooking, calling the
method "*à l'anglaise,*" which we know to be an insult, given what the
French think of English food. One French scientist—Hervé This—
goes so far as to accuse the method of "intellectual poverty." French
cooks are fond instead of braising vegetables such as carrots in a
tiny amount of water with butter, or stewing them like ratatouille, or
baking them with stock or cream in a gratin to concentrate their
sweetness; boiling is—perhaps rightly—regarded as the dullest way.

As a form of technology, however, boiling is far from obvious.
The pot transformed the possibilities of cooking. To be able to boil
something—in a liquid, which may or may not impart additional flavor—was a big step up from fire alone. It's hard to imagine a kitchen
without pots and therefore hard to appreciate how many dishes we

owe to this basic form of equipment. Pots enabled consumption of a far wider range of foods: many plants that had previously been toxic or at least indigestible became edible, once they could be boiled for several hours. Pots mark the leap from mere heating to cuisine: to the calm, considered intermingling of ingredients in a man-made vessel. Historically, the earliest cooking was roasting or barbecuing. Evidence of roasting goes back hundreds of thousands of years. By contrast, clay cooking pots date back only around 9,000 or 10,000 years. Large stone cooking pots from the Tehuaca Valley in Central America have also been found from sometime around 7000 BC.

Roasting is a direct and unequivocal form of cooking: raw food meets flame and transforms. Boiling and frying are indirect forms. In addition to fire, they require a waterproof and fireproof vessel. The food only takes on the heat of the fire through a medium, whether oil for frying or water for boiling. This is an advance on crude fire, particularly when cooking something delicate such as an egg. When you boil an egg, it is removed from the onslaught of the fire by three things: its own shell, the wall of the cooking vessel, and the bubbling water. But boiling water is not something encountered in nature very often.

Geothermal springs can be found in Iceland, Japan, and New Zealand. They are sufficiently rare, however, that they still have the status of a natural wonder. In preindustrial times, living near hot springs must have been like having a samovar the size of a lake in your backyard: an improbable luxury. The Maori of New Zealand who lived close to the boiling pools of Whakarewarewa traditionally used them for cooking. Food of various kinds—root vegetables, meats—would be placed in flaxen bags and suspended in the water until cooked. A similar technique has been practiced in the geothermal regions of Iceland for hundreds of years. Today in Iceland, a kind of dark rye bread is still made by placing the dough inside a tin and burying it in the hot earth near to the springs until it is fully steamed (which usually takes around twenty-four hours).

The archaeological evidence is unclear, but it is reasonable to assume that ancient peoples living near geysers experimented for many thousands of years with dipping raw foods into the swirling steam, attached to a stick or string that could be used to whip out the food once it was done. Ideally. Unless our ancestors were far more dexterous than we are, many pieces of perfectly good food would have gotten lost in the volcanic water, like chunks of bread tumbling into a fondue pot.

Still, geyser cooking has many advantages over fire cooking. It is less labor-intensive—all the work of creating a heat source is avoided. It is also gentler on the ingredients themselves. When cooking directly in the fire, it is hard to avoid the problem of charred on the outside and raw in the middle. Food bathed in hot water, on the other hand, can cook in its own good time; a few minutes more or less do not desperately matter.

Most people, however, do not live anywhere near geothermal springs. If you had only encountered cold water, what would it take for you to get the idea of heating it up to cook with? Water and fire are opposites; enemies, even. If you had spent hours getting your fire going—the wood gathering, the flint rubbing, the piling up of sticks—why would you jeopardize it all by bringing water near your precious hearth? To us, with our easily reignitable burners and electric kettles, boiling is a very prosaic activity. We are accustomed to pots. But cooking in hot water would not have seemed the obvious next step to someone who had never done it.

The first conscious acts of boiling took great invention. To make a vessel for cooking when there was none before is a feat of huge creativity. In geothermal cooking, although various bags and strings may be used, they are not essential: the earth itself containing the bubbling water becomes the cooking pot. In the absence of hot springs, however, boiling requires a container, one strong enough to withstand heat and from which the food will not leak. In the days before clay pots, what could it be?

Before the first potter fashioned the first pot, certain foods came ready to cook in their own vessels. Shellfish and various reptiles, notably turtles, seem to have their own pottery casing. Seashells are still used as serving vessels and utensils. When you eat a steaming bowl of moules marinières, you first choose one of the mussels as a handy pair of tongs to pick out the flesh from the other mussels. Similarly, the indigenous early Yahgan people of Tierra del Fuego used mussel shells as a dripping pan, to catch the grease from a seal as it roasted.

Several anthropologists have suggested that it would have been a small step from using mussel shells in such a way to cooking in containers. Shells have often been spoken of as one stage on the route to man-made pots. But were they?

A mussel is hardly big enough to boil or fry anything in but itself. Catching drips of fat is more the action of a spoon than of a pot. Native Americans were among those who used clam shells for spoons and sharpened mussel shells as knives for carving fish; but they did not use them for pots, so far as we know. A pearly mussel pot—it's an appealing thought—would only be large enough for dinner to feed a mouse. What, though, of larger mollusks, and reptiles? It has been said that the example of turtle cookery—as practiced by various Amazonian tribes—proves that boiling was "viable" long before the invention of pottery. Cooking in a turtle shell is certainly a romantic notion. Whether anything was cooked in turtle shells except for turtles themselves is another matter.

Moving on from shells, there are some more plausible candidates for the first cooking vessels. Tough-rinded vegetable gourds of various kinds made very handy prehistoric bowls, bottles, and pots. Hollowed-out bamboo stems, used all over Asia, are another plant-based family of cooking vessels. But bamboo and gourds were only to be found in certain parts of the world. A more universal vessel, after the discovery that meat could be cooked, was the animal's stomach, a premade container that was both waterproof and—up to a point—heatproof. Haggis, beloved of the Scots, boiled in a sheep's stomach, is a throwback to the ancient tradition of boiling the contents of an animal's belly in the stomach itself. In the fifth century

BC, the historian Herodotus recounted how the nomad Scythians used this technique, boiling an animal's flesh inside its own paunch: "In this way an ox, or any other sacrificial beast, is ingeniously made to boil itself." Ingenious is the word. The tradition of stomach cookery shows how sharp-witted humans were in finding better methods to cook their dinner, when they had no pots and pans, no Teflon nonstick griddles, no gleaming copper *batterie de cuisine* neatly dangling from pot hooks.

No method was as ingenious as the technology of hot-stone cookery practiced across the globe, starting at least 30,000 years ago. After thousands of years of direct-fire roasting, people finally figured out a more *indirect* way of using heat to cook things in steam or water. It has been said that this transformation in how food could be cooked was the greatest technological innovation in food preparation until modern times.

This is how to make a pit oven. First, dig a large hole in the ground and line it with stones to make it roughly waterproof. Then, fill the pit with water. You could skip this stage if you dug the pit below the water table, in which case it would fill up automatically. (In Ireland, there are thousands of traces of hot-rock troughs cut into the watery peat bog.) Next, take some more stones—preferably, large river cobblestones—and heat them to a very high temperature in a fire. Cooking rocks were heated as hot as 932°F, hotter than a pizza oven. Transport the stones to the pit, using tools such as wooden tongs to avoid burning your hands, and drop them in the water. When enough stones have been added, the water will start to "seethe" or boil and food can be added, topped with an insulating lid of turf, leaves, animal skins, or earth. As the temperature of the water drops, continue to add more hot rocks to keep the boiling constant until the meal is cooked.

There were many variations on stone cookery. Sometimes the stones were heated up inside the pit itself instead of in a separate fire; there would be two adjacent sections, one for the water, one for the fire and the rocks. Sometimes the food was steamed instead of boiled. Root vegetables or pieces of meat could be wrapped in leaves and layered up in the pit with the hot stones without added water, in which case the earth pit was more like an oven than a boiler.

Hot-rock cookery is still practiced in the clambakes of New England, in which sweet clams, just harvested, are cooked right there on the beach, layered up in a pit of hot stones, driftwood, and seaweed, which keep the clams juicy. The method is also used in the Hawaiian luau, in which a pig is covered in banana or taro leaves and buried in a hot pit (an *imu*) for the best part of a day, then unearthed with great ceremony and jubilation. In the Old World, however, rock boiling did not live long after the beginnings of pottery.

It is easy to assume, therefore, that cooking with stones is simply an inferior technology, compared to boiling something in a pot. But is it? It is certainly an inconvenient and roundabout way of making a hot meal. Pit boiling would be a hopeless method for doing the kind of boiling most of us do routinely: pasta, potatoes, or rice would get lost in the mud, and it would be an absurdly inefficient way of boiling things like eggs or asparagus spears, which only take a few minutes.

Hot-stone cooking was a superb technology, however, for many of the uses to which it was actually put by cooks of the past. It was great for cooking foods in bulk, as the example of the luau pig demonstrates. The other notable thing about pit-stone cookery was that it made it possible to eat numerous wild plants that would otherwise have been more or less inedible. The types of foods traditionally cooked in the slow, moist heat of a pit oven tended to be bulbs and tuberous roots rich in inulin, a carbohydrate that cannot be digested by the human stomach (it is present in Jerusalem artichokes, hence their notorious flatulent effects). Hot-stone cookery transformed these plants through hydrolysis, a process liberating the digestible fructose from the carbohydrate. In some cases, these plants needed to be cooked for as long as sixty hours for the hydrolysis to occur. A pleasant side effect was that the long, moist cooking made unpromising wild bulbs taste fantastically sweet.

Some people were so attached to earth ovens and pit boiling that they did not see pots as superior or even necessary. The Polynesians of the early Christian era—the people who traveled to the eastern Pacific islands in the first millennium AD, arriving in Hawaii, New Zealand, and Easter Island from Samoa and Tonga—present the fascinating spectacle of people who had known pots for a thousand years, only to abandon them. From around 800 BC, Polynesians made a range of pottery, typically earthenware fired at low heat, tempered with shell or sand. Yet when they arrived in the Marquesas Islands, around 100 AD, they abruptly gave up pottery making and chose to cook once again without pots. Why?

The hypothesis used to be that the reason Polynesians stopped making pots was that their new island homes lacked clay. But this was not so; clay was present on the islands, albeit in rather remote high places. Thirty years ago, the New Zealand anthropologist Helen M. Leach suggested a radical new explanation for the Polynesian conundrum: they cooked without pots because they did not see the need for them. It might have been different if they had been rice eaters. But the Polynesian diet was rich in starchy vegetables such as yams, taro, sweet potato, and breadfruit, all of which cooked better with hot stones than in pots.

So, yes, it is possible to boil without pots. The Polynesian rejection of pottery is a useful reminder that even the most basic-seeming of kitchen technologies are not universally adopted. Some cooks refuse to have a frying pan in the house (as if its very presence might cause you to consume unhealthy amounts of fat); raw foodists reject the use of fire; and there is probably someone, somewhere, who chooses to cook without knives; certainly, there are children's cookbooks that advocate the use of scissors instead. I myself am the opposite of a Polynesian. I view pots and pans as essential kitchen furniture, unassuming household gods. Few moments in the day are happier than when I sling a pot on the stove, knowing that supper will soon be bubbling away, filling the house with good scents. I can't imagine living without them.

Once pots were embedded as a technology, we developed strong feelings about them. Pottery is deeply personal. Even now, we describe pots as having human characteristics. Pots may have lips and mouths, necks and shoulders, bellies and bottoms. The Dowayo people of Cameroon in Africa have special forms of pottery for different people (a child's bowl would look different from one belonging to a widow), and there are taboos against eating from another person's designated food pot.

Many of us cling to particular vessels, fetishizing over this mug or that plate. I do not care what fork I eat with, or if anyone else has eaten with it before me (so long as it is reasonably clean). Pottery is different. I used to have a large mug with all the American presidents on it that my husband brought back from a trip to Washington. It was what I drank my early morning tea out of. The tea didn't taste the same from any other mug; it was a crucial part of the morning ritual. Gradually, the faces of the presidents faded and it was hard to distinguish Chester Arthur from Grover Cleveland. I loved it all the more. If I saw someone else drinking from it, I secretly felt that they were walking on my grave. Eventually, the mug smashed in the dishwasher, which was a relief in a way. I didn't replace it.

Fragments or shards of ceramics are often the most durable traces left by a civilization, offering our best window on the values of those who used them. Archaeologists therefore like to name people after the pots they left behind. There are the Beaker folk of the third millennium BC, who traveled across Europe, from the Spanish Peninsula and central Germany, reaching Britain around 2000 BC. They came after the Funnelbeaker culture and the Corded Ware people. Wherever they went, the Beaker folk left traces of reddish-brown, bell-shaped clay drinking vessels. They could have been named the Flint Dagger people or the Stone Hammer people (because they also used these) but somehow pottery is more evocative of a whole culture. We know that the Beaker folk liked to be buried with a beaker at their feet, presumably for the food and drink they would need in the afterlife. Our own culture has so much *stuff* that pottery has lost much of its former importance, but it is still one of the few universal possessions. Perhaps many hundreds of years from now, when our culture has been buried by some apocalypse or other, archaeologists will start to dig up our remains and name us the Mug community, MC for short: we were a people who liked our ceramics to be brightly colored, large enough to accommodate high volumes of comforting caffeinated drinks and above all dishwasher-proof.

The very existence of pottery marks a supremely important technological stage in the development of human culture. The potter takes sloppy, formless clay, wets it, tempers it, molds it, and fires it, and so gives it durable shape: this is a different order of creation from chipping away at rock or wood or bone. Clay pots bear the marks of human hands. There is a kind of magic to the process of pottery, and indeed, early potters often had a second role as shamans in the community. The archaeologist Kathleen Kenyon, who unearthed numerous pottery shards at Jericho, dating back to 7000 BC, described the beginnings of pottery as an "industrial revolution":

> Man, instead of simply fashioning an artifact out of natural material, has discovered that he can alter some of these materials. By making a mixture of clay, grit, and straw and subjecting it to high temperature, he has actually altered the nature of his material and given it new properties.

Making a usable pot is not just a matter of lumping wet clay into the relevant shape, like making a mud pie. The clay itself has to be carefully selected (too much grit and it won't form easily; not enough grit and it won't stand up to firing). The potter (who would often have been a woman) knows how to use just enough water to make the clay slippery, but not so much that the wet clay slides apart in her hands or cracks in the fire. The fire itself must be scorching hot—maybe 1652°F to 1852°F—something that can only be achieved with a custom-built kiln oven. As for making pots specifically for cooking, this is even harder, because they need to be both water-tight and strong enough to withstand thermal shock: in a poorly made pot, different materials expand at different rates as it heats up and the stress causes it to shatter.

Most cooks experience thermal shock at one time or other: the dish of lasagna that unexpectedly snaps in a hot oven, ruining your dinner plans; the supposedly "flameproof" earthenware bean pot that shatters on the stove, disgorging its contents on the floor. Food writer Nigel Slater observes that it is preferable for a pot to "shatter into a hundred pieces than sustain a deep crack. The Cracked Pot might still be a favourite, but it introduces an element of danger I can live without . . . that uneasy feeling when you open the oven door that the dish will be in two halves, macaroni cheese sizzling on the oven floor."

We will never know exactly how the first pot was made. Pottery is one of those brilliant advances that curiously occurred to different people simultaneously in far-flung places. Pots suddenly become common around 10,000 BC, or a bit before, in South America and North Africa, and among the Jomon people of Japan. The Japanese word *Jomon* means "cord-marked." Jomon pottery shows what artistry went into ceramics from a very early date. It wasn't enough to make a good pot; it had to be beautiful. Having formed their pots, Jomon potters decorated the wet clay with cords and knotted cords, with bamboo sticks, with shells. Most of the very earliest Jomon pots seem to have been used for cooking: the surviving shards indicate deep, round-bottomed flowerpot-shaped pots, ideal for stewing.

Strangely, the Jomon adoption of pots for food was not echoed everywhere. It used to be assumed that people started to make pots specifically for the purpose of cooking. But now there are doubts. How can we know whether people cooked with pots or not? Fragments of cooking pots will bear signs of scorching or mottling from exposure to the fire; they may even contain traces of food; and they are likely to be made from heavily tempered or gritted clay, fired low to eliminate thermal shock.

In the Peloponnese in Greece there is a cave called the Franchti, from which more than 1 million pottery shards have been recovered, dating from 6000 to 3000 BC. This is one of the oldest agricultural sites in Greece. People here farmed lentils, almonds and pistachios, oats and barley. They ate fish. In other words, here were people who could really use some cooking pots. One might assume that those pottery fragments once belonged to cooking pots and storage jars. Yet when archaeologists examined the oldest fragments at Franchti, they found that they bore none of the telltale signs of being held over a fire. They were not sooty or charred, but highly burnished, glossy, fine ware, made in angular shapes that would not sit well on a fire. All the signs were that these pots were used not for food but for some kind of religious ritual. This is a puzzle. These Greek settlers had at their disposal all the technology they needed to make cooking pots, but they chose not to, preferring to put their clay to symbolic use. Why? Probably because no one there had ever used pots for cooking in the past, so it just did not occur to them to do so in this later era.

Cooking pots represented a huge innovation. It took many hundreds of years of using pots as decorative or symbolic objects for the Greeks at Franchti to think of cooking in them. It is only among the later fragments, toward 3000 BC, that cooking ware becomes the norm. The Franchti pots become rounded and coarser in texture and are made in a variety of handy shapes for different tasks: stew pots of various sizes, cheese pots, clay sieves, and larger pots in ovenlike shapes. At last, these people had discovered the joys of cooking with pots and pans.

The Greeks are perhaps the most celebrated of all potters. It's easy to focus on the archetypal red-on-black and black-on-red show pots, depicting battle scenes and myths, horsemen, dancers, and feasts. But we can learn just as much from their plainer cooking pots, whose story is less dramatic but no less interesting. Greek kitchen pots tell us what they ate and how they ate it, which foods they prized and what they did with them. The Greeks left behind numerous storage jars: for cheese and olives, for wine, for oil, but above all for cereal, most likely barley: sturdy terracotta bins with lids to keep out the insects. Greek potters made frying pans, saucepans, and casseroles from coarse, gritty clay: the basic shape was the round, amphora-like *chytra*. They made little three-legged pots and handy combination-sets of casseroles and braziers, with the vessel and the heater designed in tandem. These were people who had more than one cooking strategy available to them.

Pottery changed the nature of cooking in radical ways. Unlike baskets, gourds, and coconut shells—or any of the other food containers used before—clay could be formed into any size or shape desired. Clay vessels hugely expanded the range of food that could be eaten. To sum it up in one word: porridge. With clay pots, cooks could easily boil up small grains, such as wheat, maize, and rice, the starchy staples that would soon form the mainstay of the human diet the world over. Pots thus worked in tandem with the new science of agriculture (which also emerged around 10,000 years ago) to change our diet forever. We went from a hunter-gatherer regime of meat, nuts, and seeds to a peasant diet of mushy grain with something on the side. This is a revolution whose effects we are still living with today. When we find our largest pot and boil up a pan of slippery spaghetti, or idly switch on the rice cooker, or stir butter and parmesan into a soothing dish of polenta, we are communing with those first farmers who learned how to fill their bellies with something soft and starchy, deliberately grown in a field and cooked in a pot.

In many cases, the clay pot enabled people to eat plants that would otherwise be toxic. An example is cassava (also known as manioc or yuca), a starchy tuber native to South America, which is now the third-largest source of edible carbohydrate in the world. In its natural form, cassava contains small amounts of cyanide. When inadequately cooked or eaten raw, it can cause a disease called konzo, a paralytic disorder. Once it was possible to boil cassava in a pot, it went from useless toxin to valuable staple, a sweet fleshy source of calcium, phosphorus, and vitamin C (though little protein). Boiled cassava is a basic source of energy in Nigeria, Sierra Leone, and Ghana, among other countries, usually eaten simply by mashing the boiled root to a comforting paste, perhaps with a few spices. This is classic pot-cooked food: the kind that warms the belly and soothes the heart.

Casseroles are a pleasure to eat largely because of the juices: that heady intermingling of herbs and wine and stock. Right from the start, pots enabled cooks to capture juices that would otherwise be lost in the flames. Pots seem to have been especially valued among people who ate a lot of shellfish, because the clay caught the luscious clam liquor. Pottery is a great breakthrough for another reason: it is much harder to burn food than when it is cooked directly in the fire (though still not impossible, as many of us can testify). So long as the pot is not allowed to run dry, the food won't char.

The earliest recipes on record come from Mesopotamia (the site of modern-day Iraq, Iran, and Syria). They are written in cuneiform on three stone tablets, approximately 4,000 years old, offering a tantalizing glimpse of how the Mesopotamians might have cooked. The vast majority of the recipes are for pot cooking, most of them for broths and court bouillons. "Assemble all the ingredients in the pot" is a frequent instruction. Pots made cooking a refined and subtle business for the first time; but pot cookery is also easier than direct-fire roasting. It was little trouble to boil up mutton and water and mash in some leeks, garlic, and green herbs, then leave it to bubble away in its own good time. The elementary pattern these Mesopotamian recipes took was: prepare water, add fat and salt to taste; add meat, leeks, and garlic; cook in the pot; maybe add fresh coriander or mint; and serve.

A whole range of techniques opened up with pottery. Boiling was the most important, but it also became possible to use ceramic griddles to cook thin maize cakes, cassava cakes, and flatbreads; to use large pots to brew and distill alcoholic drinks; and to use a dry, lidded pot to toast grains, the most notable example of this being the popped maize of Mesoamerica: popcorn!

People loved clay pots for another reason: the way they made the food taste. In modern times, we have more or less discarded the idea of a pot's surface mingling with its contents. We want pots to be made from surfaces that react as little as possible with what is inside: this is one of the many virtues of stainless steel. With a few theatrical exceptions—the 1970s chicken brick, the Thai claypot—we do not consider the possibility that the cooking surface could react with the food in beneficial ways. But traditionally, cultures that cook with porous clay appreciate the flavor it gives to the food, a result of the free soluble salts in the clay leaching out. In the Kathmandu Valley in India, a clay pot is considered essential for pickle jars, adding something extra to mango, lemon, and cucumber pickles.

Clay's special properties may explain why many cooks resisted the next great leap forward: the move from clay pots to metal pots. Metal cauldrons are a product of the Bronze Age (circa 3000 BC onward), a period of rapid technological change. They belong to roughly the same era as early writing systems (hieroglyphics and cuneiform), papyrus, plumbing, glassmaking, and the wheel. Cauldrons started to be used by the Egyptians, the Mesopotamians, and the Chinese, by at least 2000 BC. The expense of manufacturing them meant that their use was limited at first to special feasts or the food of the afterlife.

Metal cauldrons have a number of highly practical advantages over pottery. A cauldron can be scrubbed clean with sand or ash, unlike unglazed earthenware, which tends to hold the residue of the previous meal in its pores. Metal conducts heat better than clay, and therefore food cooks more efficiently. Most significant, a cauldron can be placed directly over fire without fear that it will shatter from thermal shock or get chipped. It might even survive dropping. Whereas archaeologists tend to encounter clay pots in the form of shards, they sometimes unearth cauldrons in their entirety, such as the Battersea cauldron in the British Museum, a splendid Iron Age specimen from 800–700 BC, which was pulled out of the River Thames in the nineteenth century. It is a magnificent pumpkin-shaped vessel, constructed from seven sheets of bronze riveted together like a shield, that has survived in all its glory. It is an awe-inspiring piece of equipment. Looking at it, you can see why cauldrons were often passed on in wills; they were weighty pieces of engineering.

Once metal cookware was possible, it wasn't long before all the basic pots and pans were established. The Romans had a patella—a metal pan for shallow-frying fish that gave its name to the Spanish paella and the Italian padella—little different from our frying pans. The ability to boil things in oil—which is really what frying is—added yet another dimension to kitchen life. Fats reach much higher temperatures than water, and food cooks quicker in oil than water, browning deliciously at the edges. This is the result of the Maillard re-

action, an interaction between proteins and sugars at high heats that is responsible for many of the flavors we find most seductive: the golden crust on a French fry, a dark spoonful of maple syrup. A frying pan is a good thing to have around.

The Romans also had beautifully made metal colanders and bronze chafing dishes, flattish metal patinae, vast cauldrons of brass and bronze, pastry molds in varying ornate shapes, fish kettles, frying pans with special pouring lips to dispense the sauce and handles that folded up. Much of what has remained looks disconcertingly modern. The range of Roman metal cookware was still impressing the chef Alexis Soyer in 1853. Soyer was particularly taken with a very high-tech sounding two-tiered vessel called the *authepsa* (the name means "self-boiling"). Like a modern steamer, it came in two layers, made of Corinthian brass. The top compartment, said Soyer, could be used for gently cooking "light delicacies destined for dessert." It was a highly valued utensil. Cicero describes one *authepsa* being sold at auction for such a high price that bystanders assumed the thing being sold was an entire farm.

Technologically speaking, Roman metal utensils have had few rivals until the late twentieth century with the advent of pans made from multilayered metals. They even addressed themselves to the problem of avoiding hot spots when cooking, which remains a bugbear for saucepan designers. A metal pan has survived from Roman Britain with concentric rings in its base, whose purpose, it seems, was to create slow, steady heat distribution. Experiments with corrugated cooking pots versus smooth ones have shown that texturing the bottom of a pan reduces thermal stresses (the rings make it less susceptible to warping over high heat, strengthening the pan's structure)—and also gives more cooking control: heat transfer happens more slowly with a textured pan, so there is less chance of annoying boilovers. A similar pattern of concentric rings appears on the base of Circulon cookware, launched in 1985, whose "unique Hi-Lo" grooves are said to reduce the surface abrasion and enhance the durability and nonstick qualities of the pan. As with aqueducts, straight roads, arched bridges, and books, this was a technology in which the Romans got there first.

Despite the ingenuity of the Romans, most domestic cooks from the Bronze Age until the eighteenth century had to make do with a single big pot: the cauldron (often called a kettle or kittle). It was by far the largest utensil in the Northern European kitchen, and the one around which culinary activity was focused. Once the Romans had fallen, the range of cookware shrank back to basics. From a pot for every occasion, the one-pot meal was once again the dominant mode of cooking. The cauldron tended to decide for you how you could eat. Boiled, stewed, or braised was usually the answer (though a covered pot could also be used to make bread, which baked or steamed under the lid). The contents of the cauldron could be fairly repetitive: "Pease porridge hot / Pease porridge cold / Pease porridge in the pot / Nine days old," as the rhyme goes. A typical modest medieval household owned a knife, a ladle, an earthenware pan, perhaps a spit of some sort, and a cauldron. The knife chopped the ingredients to go in the cauldron along with water. Several hours later, the ladle poured out the finished soup or "pottage." Supplementary pots took the form of a few cheap earthenware pots, and perhaps a skillet, which is a long-handled pan much smaller than the cauldron, used for heating up milk and cream.

If further kitchen tools were owned, they were most likely accessories to the cauldron. Iron pot cranes or sways, some of them beautifully ornate, were designed to swing the heavy pot and its contents on its hook over the fire and off again, a form of temperature control as instant as flicking a switch, if rather more dangerous. Those who could not afford such elaborate machinery might own a brandreth or two, ingenious little three-legged stands designed to lift the cauldron above the direct heat of the fire. Flesh-hooks and flesh-forks were other cauldron accessories, used for suspending meat over the bubbling liquid or for retrieving things from its depths.

Cauldrons came in many shapes and sizes. In Britain, they were usually "sag-bottomed" (as opposed to pot-bellied) and made of bronze or iron, so that they could withstand the heat of the fire. If they had three legs, this was a sign they were designed to sit in the embers. Iron cooking pots, which tended to be smaller, were round-bellied, with handles for hanging over the fire. Sticks or tongs were used to manipulate the handle, which would become prohibitively hot. Cooking with a single pot could give rise to strange combinations of ingredients, all jumbled together. It is not clear how often the cauldron was cleaned out, in the absence of running water and dish detergent. Mostly, the scrapings of the previous meal were left in the bottom to season the next one.

European folklore is haunted by the specter of the empty cauldron. It is the old equivalent of the empty fridge: a symbol of outright hunger. In Celtic myth, cauldrons are capable of summoning up both eternal abundance and absolute knowledge. To have a pot and nothing to put in it was the depths of misery. In the story called "Stone Soup" (which has many versions) some travelers come to a village carrying an empty cooking pot and beg for some food. The villagers refuse. The travelers produce a stone and some water and claim they are making "stone soup." The villagers are so fascinated, they each add a little something to the pot—a few vegetables, some seasoning—until finally the "stone soup" has become a rich cassoulet-like hot pot, from which all can feast.

Acquiring a cauldron required a sizable outlay. In 1412, the worldly goods of Londoners John Cole and his wife, Juliana, included a sixteen-pound cauldron worth four shillings (the cost of an earthenware pot at this time was around a penny, with twelve pennies to the shilling). Once bought or bartered, a metal pot might be repaired many times to prolong its life; if it sprang holes, you would pay a tinker to solder them. A bronze cauldron was dug up in County Down in a bog in 1857. It showed six areas of repair; small holes had been filled in with rivets; larger ones were mended by pouring molten bronze onto the gap.

A cauldron might not be the ideal vessel for every dish. But once acquired, it was likely to be the one and only pot (supplemented, if at all, by a small earthenware vessel or two). Every culture has its own take on the one-pot dish, as well as variations on the specific pot that was used to make it: pot au feu, Irish stew, dobrada, cocido. One-pot cookery is a cuisine of scarcity: scarce fuel, scarce utensils, scarce ingredients. Nothing is wasted. It is no coincidence that food for the relief of poverty has almost always taken the form of soup. If there is not enough to go around, you can always add some more water and bubble it up one more time.

Cooks devised some crafty ways around the limitation of the single pot. By putting vegetables, potatoes, and pudding in separate muslin bags in the boiling water, it was possible to cook more than one thing at once in a single vessel. The pudding might end up tasting a bit cabbagey, and the cabbage rather puddingy, but at least it made a change from soup. In *Lark Rise to Candleford*, Flora Thompson describes how "tea" was made for the men coming home from the fields:

> Everything was cooked in the one utensil; the square of bacon, amounting to little more than a taste each; cabbage, or other green vegetables in one net, potatoes in another, and the roly-poly swathed in a cloth. It sounds a haphazard method in these days of gas and electric cookers; but it answered its purpose, for, by carefully timing the putting in of each item and keeping the simmering of the pot well regulated, each item was kept intact and an appetising meal was produced.

In the 1930s, the Nazis borrowed the frugal image of the one-pot meal, putting it to ideological use. In 1933, Hitler's government announced that Germans should put aside one Sunday, from October to March, to eat a one-pot meal: Eintopf. The idea was that people would save enough money in this way to donate whatever was saved to the poor. Cookbooks were hastily rewritten to take account of the new policy. One recipe collection listed no fewer than sixty-nine Eintopfs, including macaroni, goulash, Irish stew, Serbian rice soup, numerous cabbagey medleys, and Old German potato soup.

The Nazi promotion of the Eintopf was a shrewd piece of propaganda. Many in Germany already viewed the Eintopf as the ultimate frugal meal, a dish of sacrifice and suffering. It was said that Germany had managed to beat the French in 1870 in part because the armies had filled their bellies with Erbswurst, a one-pot mixture of pea meal and beef fat, a kind of pease pudding. The Eintopf came with a sea of nostalgic memories.

The Nazis' celebration of the Eintopf was actually a sign of how most kitchens—in Germany, as elsewhere—had moved beyond one-pot cookery. Like many other fascist symbols, it harks back to the archaic. You could only see the Eintopf as a money saver in a society in which most meals were cooked using more than one pot. By reviving the fairy-tale peasant ideal of a single cauldron hung on a single pothook, the Nazis inadvertently showed that the days of the cauldron were over. Even though times were tough in 1930s Germany, most cooks—which meant most housewives—expected to have an assortment of pots and pans to cook with, not just one.

Petworth House in Sussex is one of the grandest residences in England. It has descended through the same aristocratic family, the Egremonts, since 1150, though the current building dates to the seventeenth century: a stupendous mansion set in a seven-hundred-acre deer park. It is now managed by the National Trust. Visitors to the kitchen can marvel at the gleaming copper *batterie de cuisine* on display, more than 1,000 pieces in all: rows of saucepans and stewpans, plus multiple matching lids, all immaculately lined up, from large to small, from left to right, on several vast dressers. The kitchen at Petworth gives you a sense of what it meant to have "a place for everything and everything in its place," as the cookery writer Mrs. Beeton said. The Petworth cooks would have had exactly the right pot for cooking each dish.

The equipment at Petworth includes stockpots with taps at the bottom to release hot water (like tea urns); multiple stewpans, sauté pans, and omelette pans in every size you could wish for; a large braising pan, with an indented lid designed to hold hot embers, so that the food was cooked from above and below at the same time. The pans devoted to fish cookery are a world unto themselves. In the grand old days, there would have been excellent fish from the Sussex coast, and the Petworth cooks were expected to do it justice. The house's kitchens contain not just fish kettles (with pierced draining plates inside so that a fish could be lifted from its poaching water without disintegrating) and a fish fryer (a round open pan with a wire drainer) but a special turbot pan (diamond shaped to mimic the shape of the fish) and several smaller pans specifically for cooking mackerel.

The kitchen at Petworth was not always so well-equipped. Historian Peter Brears studied inventories of the kitchen, documenting "every single movable item" used by the cooks; every pot, every pan. The first inventory took place in 1632; then 1764; then 1869. These documents offer a snapshot, century after century, of what cooking equipment was available in the richest British kitchens. The most telling detail is this. In 1632, during Stuart times, for all its wealth, Petworth owned not a single stewpan or saucepan. The devices for stewing or boiling at that time were: one large fixed "copper" (a giant vat that held boiling water, used to supply hot water for the whole house, not just for cooking); nine stockpots (or cauldrons); an iron cockle pan and a few fish kettles; and five small brass skillets, three-legged to stand in the fire. This is not a kitchen in which you could concoct an hollandaise or an espagnole sauce. You could stew, poach, or boil, but not with any great finesse. The focus of this kitchen was roasting not boiling: there were twenty-one spits, six dripping pans, three basting ladles, and five gridirons.

By 1764, this had all changed. Now, Petworth had shrugged off some of its spits (only nine remained) and acquired twenty-four large stewpans, twelve small stewpans, and nine bain-maries and saucepans. This massive increase in the number and variety of pans reflects new styles of cooking. The old heavy spicy medieval ways were on the way out, to be replaced by something fresher and more buttery. An aristocrat in 1764 was familiar with many foods that simply were not known in 1632: with frothy chocolate and crisp cookies; with the sharp, citrusy sauces and truffley ragouts of French nouvelle cuisine. New dishes called for new equipment. Hannah Glasse, one of the most celebrated cookery writers of the eighteenth century, felt it was important to get the right pan when melting butter (a kind of thickened melted butter was starting to be served as a universal sauce to go with meat or fish): a silver pan was always best, she advised.

By 1869, the Petworth kitchens had even more pans. Peter Brears suggests that Victorian cooks would have found the ample equipment of 1764 to be "totally inadequate." The focus of the kitchen was finally moving away from spit roasting. The real action was now happening in copper pans, resting on steam-heated hotplates. There were also now three steamers, for food that needed gentler water-cooking than boiling. The number of stewpans and saucepans had gone up from forty-five to ninety-six, a sign of the sheer volume and variety of different sauces, glazes, and garnishes required by Victorian cuisine.

Incidentally, what is the difference between a stewpan and a saucepan? Not much, is the answer. In the eighteenth century, saucepans tended to be smaller, suitable for the furious whisking of emulsions and glazes. They did not necessarily have a lid, because they were often used simply for warming up sauces and gravies that had already been made in a stewpan and strained through a sieve. Stewpans were bigger and lidded; they might hold multiple partridges or an assortment of ox cheeks, red wine, and carrots; a chicken fricassee or a delicate liaison of lamb's sweetbreads and asparagus. The stewpan was what got dinner on the table. Over time, however, the saucepan gained ground. In 1844, Thomas Webster, author of *An Encyclopaedia of Domestic Economy*, wrote that saucepans were "smaller round vessels for boiling, made with a single handle," whereas stewpans were made with a double handle, one on the lid and one on the pan. He added that stewpans were made of a thicker metal and tended to have a rounded, less angular bottom, which made them easier to clean. We no longer speak of stewpans, using the grand term "saucepans" for all our basic pans, lidded or otherwise, even when we use them for nothing more elevated than heating up a can of beans.

Many kitchens still allude, in a modest way, to the *batterie de cuisine*. It might be a trio of enameled pans stacked in a pot holder; or an orderly row of Le Creuset, arranged from small to large. The *batterie de cuisine* was one of many new ideas to come out of the eighteenth century, era of enlightenment and revolution. The thinking behind the *batterie* was the exact opposite of the limitations of one-pot cooking. The idea—which still has fierce believers among the practitioners of haute cuisine—is this: every component of a meal requires its own special vessel. You cannot sauté in a sloping-sided frying pan or fry in a straight-sided sauté pan. You cannot poach turbot without a turbot kettle. You need the right tool for the job. In part, this reflects the new professionalism of cooking in the eighteenth century and the influence of France.

At E. Dehillerin, the oldest surviving kitchen shop in Paris, you can still worship at the temple of copper cookware. The green-fronted shop is replete with vessels you never knew you needed: a snail dish for cooking garlic snails, molds for the most fanciful patisseries, tiny  sauce pans that really are intended for sauce making, a press for making a very specific dish of pressed duck in which the carcass is crushed until the organ juices run out, lidded ragout pans, stockpots, and yes, even a copper turbotière that looks very like the one at Petworth. This place seems to be infused with the spirit of Julia Child, who began her *Mastering the Art of French Cooking* with a stern piece of advice: do not be a pot saver. "A pot saver is a self-hampering cook. Use all the pans, bowls and equipment you need."

William Verrall was an eighteenth-century chef and the landlord of the White Hart Inn in Lewes, Sussex, who published a cookbook in 1759. Verrall had no time for those kitchens that attempted to make do with "one poor solitary stewpan" and a single frying pan "black as my hat." For Verrall, it was obvious that "a good dinner cannot be got up to look neat and pretty without proper utensils to work it in, such as neat stew-pans of several sizes" and omelette pans and soup pots. Verrall tells the tale of "half of a very grand dinner" being entirely spoiled "by misplacing only one stewpan."

This new fussiness about pans, from the eighteenth century on-ward, was fueled by a resurgence in the English copper industry. Previously, supplies of copper had been imported from Sweden. In 1689, however, the Swedish monopoly was canceled and English copper began to be produced—much of it in Bristol—in larger quantities and at a much lower cost. This paved the way for dressers heaving with copper pans. The French term *batterie de cuisine*—which became the universal way to refer to cooking equipment from the early nineteenth century on—harks back to copper pans. A *batterie* was copper that had literally been battered into shape.

The Victorian copper *batterie* is in its way the high point of the long history of pots and pans. The combination of craftsmanship, the quality of the metal itself, a preparedness to tailor the equipment to the requirements of cooking, and the existence of wealthy kitchens equipped with the battalion of cooks needed to keep track of the various vessels, would never be equaled, unless in the twentieth-century kitchens of French haute cuisine.

It is interesting, then, that despite their fabulously well-equipped kitchens, the Victorians have a reputation for having ruined British cooking, reducing everything to a mass of brown Windsor soup. Some historians have argued that this reputation is unjustified. But there is no getting away from the question of vegetables. Victorian and Regency recipes consistently tell us to boil vegetables for many times longer than we know they need. Broccoli: twenty minutes. Asparagus: fifteen–eighteen minutes. Carrots (this one is really criminal): forty-five minutes to an hour. What good is it having state-of-the-art pans in which to boil things if you have not worked out the basic method for boiling vegetables?

It is possible, however, that the Victorians did not abuse their vegetables quite as much as we suppose. The assumption has generally been that the Victorians overcooked their vegetables because they did not give the matter enough thought. But maybe the opposite was true: they overthought it. Nineteenth-century food writers were highly sensitive both to the texture of what they were cooking—like us, they sought to cook vegetables until "tender"—and to the vigor with which they boiled things. It is true that they feared the indigestibility of undercooked vegetables—as cooks had for centuries: raw vegetables had been deemed harmful ever since the humoral medicine of the Greeks. But they no less feared spoiling vegetables by overcooking them. William Kitchiner, author of *The Cook's Oracle*, noted that when cooking asparagus, "Great care must be taken to watch the exact time of their becoming tender; take them up just at that instant, and they will have their true flavour and colour: a minute or two more boiling destroys both." These are not the words of someone who intends to produce vegetable mush. It also seems an odd thing to say, given that Kitchiner has just recommended that we boil our asparagus for twenty to thirty minutes. Then again, he ties it in a bundle, and it does take much longer that way than when boiled as individual stalks.

The long boiling times were not arrived at mindlessly. We sometimes patronizingly forget that great thought has always gone into how best to cook. Most nineteenth-century recipe writers were keen to offer advice based on "scientific" or at least "rational" evidence. The most important fact about boiling, so far as they were concerned, was that the temperature of boiling water never rose above 212°F—after that, it turns to steam, but it can never get hotter. Scientists such as Count Rumford lamented the fuel inefficiency of cooking food at a galloping boil: what was the point, when it did not raise the temperature of the water? It was just a waste of energy. In 1815, Robertson Buchanan, an expert on fuel economy, noted that once it has reached the boiling point, "water remains at the same pitch of temperature, however fiercely it boils"; cookery writers often quoted Buchanan on this point. William Kitchiner said he had experimented with placing a thermometer in water "in that state which cooks call gentle simmering; the heat was 212°F, i.e., the same degree as the strongest boiling." The logic of this was that it was best to boil things at a slow simmer. In 1868, Pierre Blot, professor of gastronomy at the New York Cooking Academy, launched an attack on those—housewives and professional cooks alike—who "abused" the art of boiling by boiling "fast instead of slowly": "Set a small ocean of water on a brisk fire and boil something in it as fast as you can, you make as much steam but do not cook faster; the degree of heat being the same as if you were boiling slowly." When it came to meat cookery, this advice to simmer slowly rather than to boil rapidly was good ("The slower it boils," said Kitchiner, "the tenderer, the plumper and whiter it will be"). But with vegetables—potatoes aside—the slow simmer was not such a boon. It resulted in vastly elongated cooking times, all the more so because cooks in possession of a fine *batterie de cuisine* were inclined to boil food in the smallest pan possible. Here is Kitchiner again:

> The size of the boiling-pots should be adapted to what they are to contain: the larger the saucepan the more room it takes upon the fire, and a larger quantity of water requires a proportionate increase of fire to boil it.

Kitchiner then quotes the maxim "A little pot is soon hot." This saying, popular in Victorian times, is certainly true. But a little pan filled with a small amount of slowly simmering water takes far longer to cook carrots than a biggish pan of properly boiling water. The advantage of having only one or two large pans instead of a panoply of every size is that you do not have the option of fitting the food to the pot. You have to give it lots of room. The worst of all worlds are those kitchens with only a few pans, all of them too small, so that when you add food to the pan it takes an age before it returns to the boiling point.

Nineteenth-century vegetables were probably far less overcooked than you might guess from the cooking times alone, especially when you take into account the fact that the vegetables themselves were different: modern seed varieties and growing methods tend to yield more tender plants. Victorian asparagus would have been stalkier, as a rule; greens and carrots would have been tougher. Even with our tender modern vegetables, the Victorian method of boiling does not result in total sogginess. I've tried slowly simmering sliced carrots crammed in a little pan for forty-five minutes. Amazingly, they still have some bite to them, though not as much as when they are thrown into a large stainless steel pan of water at a rolling boil for five minutes, or, better still, steamed in a steamer.

The Victorian mastery of boiling technology was flawed. It's perfectly right that at normal pressure you can never get water hotter than 212°F (at higher pressures, it can get much hotter, which is why a pressure cooker cooks food so fast). But this is not the only factor determining how fast food boils. Also important is ebullition—the extent to which boiling water bubbles. In basic terms, heat transfer in cooking is determined by the difference in temperature between the food and the source of heat. On paper, therefore, the Victorian logic looks sound: once you have gotten cooking water at or near 212°F, it shouldn't really make much of a difference whether the water is vigorously bubbling or only simmering. Yet our eyes and taste buds tell us that it does. The reason is that properly boiling water moves chaotically and transfers heat to the food several times faster than simmering water. The heat transfer also works quicker when there is more water in the pan in proportion to the food. A large pan with plenty of water and not too many vegetables in it will cook far faster than a perfectly tailored little copper pan crammed to capacity. This explains why even when Victorians advise boiling vegetables "briskly," as Isabella Beeton sometimes does in *Mrs. Beeton's Book of Household Management*, the cooking times are still long.

We of the pasta generation know this instinctively. We may not be able to rustle up a meat glaze or a charlotte russe. If you gave us a copper turbot kettle, we would have no idea what to do with it, not that this matters, because the fillets of fish we mostly consume are fine when poached in a normal pan. But we mostly understand how to fast-boil far better than the Victorians: we take a package of fusilli, get out our largest pot, and boil it as fast as we can in an abundance of water for ten minutes until perfectly al dente, before tossing with butter or a rich tomatoey sauce. The single thing we look for in a pasta pot is large volume. Having mastered this skill, we can easily transfer it to vegetables: four minutes for broccoli, six for green beans, anoint with sea salt and a spritz of lemon and eat. Victorian cooks performed many feats far more daunting: jellies shaped like castles, architectural pies. But the simplicity of boiled vegetables was beyond them.

Victorian boiled food had another drawback: the pans themselves. Copper is a wonderful conductor of heat; the only pan metal more conductive is silver. But pure copper is poisonous when it comes into contact with food, particularly acids. Copper pans were thinly lined with neutral tin, but over time the surface of the tin wore down, exposing the copper beneath. "Let your pans be frequently retinned" is common advice in cookbooks of the eighteenth and nineteenth centuries. If human beings then were anything like human beings now, cooks must often have postponed retinning the pans and ended up poisoning those they cooked for. Cooks ignorant of the ill effects of copper actually sought out its greening powers, using unlined copper pans to make pickled green walnuts and green gherkins. In short, copper pans are great, apart from the fact that they potentially make food taste bad and poison you. Suddenly, those shiny Victorian *batteries de cuisine* do not look quite so desirable.

The search for the ideal cooking pot is not easy. There is always a trade-off. As the great food writer James Beard once put it: "Even in this best of all possible worlds, there is no such thing a perfect metal for pots."

We expect many things of a good pan, and not all of them are to be found in a single material. First and foremost, it should be highly conductive, so that it heats food quickly and distributes heat evenly across the base (no hot spots!). It should balance well in the hand and be light and easy to maneuver on the stove top, with a handle you can use without burning yourself. But we also want it to be dense and solid enough to withstand high heat without buckling, chipping, or cracking. The ideal pan should have a surface that is nonreactive, nonstick, noncorrosive, easy to clean, and long lasting. It should have a pretty shape and sit well on the burner. Oh, and it shouldn't cost a fortune. Over and above all this, a truly great pot has some quality—impossible to quantify—that makes it not just functional but lovable: *Hello, old friend,* you think, as you haul it out once again.

Traditionally, cookbooks started with a list of equipment required. As the author runs through the range of materials from which a pan might be constructed, there is a constant air of ambivalence, of "Yes, but . . ." Ceramic, for example, is great until it cracks. Ditto, glass ovenware or Pyrex, which is fine in the oven but fragile over a flame. Aluminum is good for omelettes, but you can't put acidic foods in it. Silver is said to be excellent except for the deluxe price tag (and the subsequent pain when it is lost or stolen); but silver-cooked foods taste of tarnish unless the pans are kept scrupulously clean. Heavy black cast-iron pans are the favorites of many cooks. Cast-iron vessels have been used for hundreds of years and are still the choice for such homely dishes as tarte tatin in France and cornbread in the United States. "Put on the skillet, put on the lead / Mamma's goin' to make a little shortnin' bread," sings Paul Robeson. If well seasoned, a cast-iron skillet has excellent nonstick properties, and because it is so heavy, it can withstand the high heat needed for searing. The downside is that these pans rust nastily if not dried and oiled carefully after use. They also leach small amounts of iron into the food (though this is a benefit if you are anemic).

The solution to many of these drawbacks was enameled cast iron: cast iron coated in a vitreous enamel glaze, the most famous example of which is Le Creuset. The principle of enameling is very ancient: the Egyptians and the Greeks made enameled jewelry, fusing powdered glass onto pottery beads by firing it at very high temperatures (1382°F to 1560°F). Enameling began to be applied to iron and steel around 1850. Then in 1925, two Belgian industrialists working in northern France thought of applying it to cast-iron cookware, the bedrock of every French grandmother's kitchen. Armand Desaegher was a cast-metal expert. Octave Aubecq knew about enameling. Together, they produced one of the definitive ranges of cookware of the twentieth century, starting with a round cocotte (we would call it a casserole) and moving over the years into ramekins and baking dishes, French ovens and tagines, roasters and woks, flan dishes and grill pans. Part of the appeal of Le Creuset cookware is the colors, which mark changing tastes in kitchen design: Flame Orange in the 1930s; Elysees Yellow in the 1950s; Blue in the 1960s (the color was suggested by Elizabeth David, inspired by a pack of Gauloises cigarettes); and Teal, Cerise, and Granite today. I have a couple in Almond (a fancy name for cream) and there is nothing better for long, slow-cooked casseroles, because the cast iron warms up evenly and retains heat superbly, while the enamel stops your stew from taking on any metallic flavors. Mostly, they score high on lovability; the sight of one on the stove makes the heart sing.

One of the best cooks I know (my mother-in-law) does all her cooking in blue Le Creuset. She was Cordon Bleu trained before she got married, and her meals have an Anglo-French panache. In her neatly kept pans, she whisks up dreamy béchamels, buttery peas, smooth purple borscht. The pans seem utterly in keeping with her style of cooking. She would never dream of serving food on cold plates or with the wrong cutlery. Her enameled cast iron serves her well. It is only when those of us with less discipline venture into the kitchen that cracks appear. For one thing, these pans are heavy, and I always fear my wrists will go limp and I'll drop one. There's also the fact that none of them is big enough for pasta. But the real trouble is the surface. If you are used to cooking on more forgiving stainless steel, it's a shock to find how easily things stick to the bottom of Le Creuset at high temperatures. Several times, I've left one of my mother-in-law's pans slightly too long on the burner and nearly ruined it (at which point she comes in and briskly saves the day with bleach).

When nonstick pans first arrived on the scene—they were first launched in France by the Tefal company in 1956—they seemed like a miracle. "The Tefal pan: the pan that really doesn't stick," was the original pitch. The reason food sticks to a pan is because proteins react with some metal ions at the surface. To prevent food from sticking, you need to stop protein molecules from reacting with the surface in this way—either by stirring it so vigilantly that it doesn't get a chance to stick, or by introducing a protective layer between the food and the pan. Traditionally, this layer is provided by "seasoning" the pan. With unenameled iron pans—whether a Chinese wok or an American cast-iron skillet—seasoning is a critical step; skip it, and your cooking will suffer (and the pan will rust). First, the pan is soaked in hot, soapy water, rinsed and dried. Then, oil or lard is rubbed into the surface and very slowly heated for several hours. Some of the fat molecules "polymerize," leaving a slick, shiny surface. Each meal that you cook adds a further layer of polymerized fat. Over time, the pan becomes as slick as Brylcreem. In a nicely blackened wok, the food slides and jumps. You can cook a whole panful of cornbread in a well-seasoned skillet, and when it is done, it will simply drop out, like a pill from a blister pack. But it takes a certain attentiveness to maintain a seasoned pan. It must never be scoured. The surface can also be ruined by acidic ingredients such as tomatoes or vinegar. When the seasoning on a cast-iron pan wears away, you have to start all over again.

In 1954, Marc Gregoire, a French engineer, came up with another way. PTFE, or polytetrafluoroethylene, had been known by chemists since 1938. The slippery substance was used for coating industrial valves and for fishing tackle. As the story goes, Marc Gregoire's wife first suggested he try to use the PTFE he had been using on fishing tackle to solve the problem of her sticky cooking pans. He found a way of melding PTFE to an aluminum pan.

How does it work? Stickiness happens when food bonds with the surface of the pan; but PTFE molecules do not bond with any other molecules. At a microscopic level, PTFE is made up of four fluorine atoms and two carbon atoms, repeated many times in a much larger molecule. Once fluorine has bonded with carbon, it does not want to bond with anything else, not even with the usual culprits such as scrambled egg or steak. Under the microscope, says scientist Robert L. Wolke, a PTFE molecule looks rather like a spiky caterpillar, and this "suit of caterpillar armor" prevents the carbon from sticking to food molecules, hence that theatrical effect when you pour a tiny bit of oil into a new nonstick pan and it seems to be repelling the droplets out of the pan.

The world went wild for Teflon. In 1961, DuPont backed the first production in the United States, called the "Happy Pan." Within the first year, American sales were 1 million units a month. Like a cure for baldness, a pan that cooks food without sticking is a universally sought-after invention. As of 2006, around 70 percent of the cookware sold in the United States has a nonstick coating; it has become the norm rather than the exception.

But as the years went on, it became obvious that nonstick was not flawless. I'd never make a stew or a sauté in nonstick, because when nonstick works, you get none of the browned sticky bits you need for deglazing. All too often, however, you have the opposite problem: the amazing nonstick properties do not last. Over time, no matter how carefully you treat it—shunning metal utensils, shielding it from searing heats—the nonstick surface of a PTFE-treated pan will simply wear away, leaving you with the metal underneath, which rather defeats the purpose. After too many short-lived nonstick pans, I've decided that it's not worth it. It's far better to buy a traditional metal like aluminum or steel or cast iron and season it with oil: that way, your pan gets better with use rather than worse. Each time you grease and cook with a cast-iron pan, it gets an extra patina. Whereas each time you cook with nonstick, the coating gets a little less slick.

There are other reasons to pause before buying nonstick pans. PTFE is a nontoxic substance, but when heated to very high temperatures (482°F and above), it emits several gaseous by-products (fluorocarbons) that can be harmful, causing flu-like symptoms ("polymer fume fever"). When doubts first emerged about the safety of nonstick pans, the industry replied that pans would never be heated this high under normal use; but by leaving a pan to preheat with no oil in it, it is perfectly possible to reach these temperatures. In addition, in 2005, the US Environmental Protection Agency looked into whether PFOA, a substance used in the manufacture of PTFE, was carcinogenic. DuPont, the main American manufacturer, has pointed out that the amount of PFOA remaining on a finished pan should not be measurable. But, whether fairly or not, many people have been left feeling uneasy about the miracle of nonstick surfaces.

Faced with all these hazards, how is one to choose the right pan? In 1988, an American engineer named Chuck Lemme, cited as the inventor on twenty-seven patents that range from hydraulics to catalytic converters, decided to approach the question systematically. He looked at all the available materials and rated them in nine categories:

1. Temperature uniformity (my translation: Will it even out heat spots?)
2. Reactivity and toxicity (Will it poison me?)
3. Hardness (Will it dent?)
4. Simple strength (Will it survive being dropped?)
5. Low stick rating (Will my dinner get glued on?)
6. Ease of maintenance (Will it wash easily?)
7. Efficiency (Does it conduct heat well vertically through the base?)
8. Weight (Can I lift it?)
9. Cost per unit (Can I afford it?)

For each category, Lemme rated the materials, using a scale of one to ten. He then tabulated his findings into an "idealness rating," with 1,000 as the perfect score.

Lemme's findings confirmed how difficult it is to produce perfect cookware. Pure aluminum rated very high for temperature uniformity (scored 8.9, out of a possible 10)—great for evenly browning an omelette—but very low for hardness (scored 2): many aluminum pans end up misshapen. Copper was efficient (scored 10) but hard to maintain (scored 1). Overall, Lemme found that none of the "single material pots" rated above 500 in the idealness scoring; in other words, they landed just halfway up the scale. The best was pure cast iron (544.4). Those of us who continue to use cast-iron skillets are on to something. But 544 is still a low score.

The only way to get closer to the ideal rating of 1,000 was to combine metals by sandwiching them together. At the time of Lemme's investigation, the consensus among high-end cookware experts was that the only copper pans worth having were fashioned from a hunk of copper as opposed to a thin, cosmetic layer. Yet Lemme found that even a very thin layer of copper "electroplated to the bottom mainly for decoration" could dramatically increase a pan's conductivity. A 1.4 mm stainless steel pan with a 0.1 mm layer of copper attached would increase its ability to even out hot spots (temperature uniformity) by 160 percent. There's a very easy way to check for hot spots in your own pans. Just sprinkle plain flour over the surface of a pan and put it over a medium-high heat. You will see a brown pattern start to form as the flour burns. If the brown patch spreads over the whole surface of the pan, you'll know that this pan has good heat uniformity. More likely, though, a small brown dot will appear toward the center: a hot spot. Now imagine that you are trying to sauté a panful of potatoes in this pan: unless you move them frequently, the ones in the middle will singe on precisely that spot while the ones at the outside remain pale. Better pans really do make a difference in the food on your plate.

Lemme's own suggestion for the "near-ideal" pot was to fabricate a composite. The inner core of the pan would be a stainless steel–nickel alloy. The inside would be coated with one of the more durable nonstick surfaces, such a flame-sprayed nickel. The outer bottom layer would be laminated with pure aluminum: 4 mm thick on the bottom, thinning out to 2 mm on the sides.

When Lemme was writing in the late 1980s, such a pan did not exist: it was a concept in the realms of sci-fi. Lemme never produced or marketed his ideal pan; it existed only in his brain, and having conceived it, he returned to other kinds of engineering. Yet even Lemme's imaginary and near-ideal pot only rated 734 on his scale. It turns out that some of the many things we want from a pan are simply incompatible. For example, a thin base makes pans more energy efficient—more quickly responsive to different heats from the burner. This can be useful for sauce making or for foods that need quick, hot cooking such as pancakes; and it results in lower energy bills. But for getting rid of hot spots, a thick metal base is better. The thickness ensures more uniform temperatures on the base of the pan and great heat retention. Thick cast iron takes ages to heat up because of its density, but once hot, it stays hot, so nothing is better for searing something like a meaty chop, because it maintains most of its heat when the cold meat hits the pan. So thin pans and thick pans are both desirable, but you can't make a pan that is thick and thin at the same time without breaking the laws of physics. Lemme's study shows that no matter how much you balance out the various factors, there will still be trade-offs. There will probably never be a pan that scores even close to 1,000 on the Lemme scale.

Nonetheless, in the intervening two decades or so, the technology of cookware has gone up a notch. As Lemme predicted, the action is all in the sandwiching together of multiple materials. All-Clad, one of the top American brands of cookware, has come up with a patented formula made of five layers of different materials, alternating higher conductive metals with lower ones to "promote the lateral flow of cooking energy and eliminate hot spots," says the company website, with a stainless-steel core to promote stability. These pans are specially designed to work with the newest-technology induction cooktops. I'm sure an All-Clad pan would score high on Lemme's scale in all ways but one: the cost runs to several hundred dollars for a single pan.

According to Dr. Nathan Myhrvold, the outlay for top-of-the-range pans may not be worth it. Myhrvold, who was the chief technology officer for Microsoft before turning to food, is the main author (along with Chris Young and Maxime Bilet) of *Modernist Cuisine* (2011), a six-volume, 2,438-page work that aspires to "reinvent cooking." Working in a state-of-the-art cooking laboratory near Seattle at his company, Intellectual Ventures (which deals in patents and inventions), Myhrvold and his team of researchers questioned the thinking behind numerous cooking techniques that had previously been taken for granted. If Myhrvold wanted to find out how food really cooks in a pressure cooker or a wok, he sliced one in half and photographed the results, midcooking. Among Myhrvold's many surprising and useful discoveries were that berries and lettuce stay fresher for longer in the fridge if you first plunge them in warm water, and that duck confit does not need to be cooked in its traditional fat—a sous-vide water bath works just as well. Myhrvold also applied himself to the problem of the ideal pan.

After extensive experiments, the author of *Modernist Cuisine* found that "no pan can be heated to perfect evenness." He noted that many (wealthy) people have expensive copper pans "hanging in a kitchen like trophies." But even the most highly conductive pan could not ensure even cooking. In all the obsessing over pots and pans, people had forgotten another basic element of the cooking process: the heat source. Myhrvold's experiments taught him that the typical small domestic gas burner, only 6 cm in diameter, was not big enough to diffuse heat evenly "to the far edges of the pan," no matter how fancy that pan might be. His advice? "Skimp on the pan, but choose your burner carefully." Assuming you have a sizable burner—ideally, as wide as the pan itself—Myhrvold found that an inexpensive aluminum–stainless steel bonded pan cooks "with nearly the same performance as that of the copper pan." Which is good to know, though not all that helpful if you are cooking in a normal, ill-equipped kitchen with average-sized burners.

There is also the question of skill. I decided to try out Myhrvold's theory on my own decidedly inferior gas burners (though at least the switches work most of the time, which is better than the stove in our old house). I took my smallest skillet and set it to heat on the largest burner to sauté some sliced zucchini. The heat conduction was appreciably more even and powerful. The discs of zucchini practically jumped out of the pan. Then they burst into flames. Since then, I have happily returned to my imperfect mishmash of too-big pans and too-small burners. I'd rather put up with the annoyance of hot spots than suffer scorched eyebrows.

The ideal pan—like the ideal home—does not exist. Never mind. Pots have never been perfect, nor do they need to be. They are not just devices for boiling and sautéing, frying and stewing. They are part of the family. We get to know their foibles and their moods. We muddle through, juggling our good pots and our not-so-good ones. And in the end, supper arrives on the table; and we eat.

~ Rice Cooker ~

WHEN ELECTRIC RICE COOKERS ARRIVED IN Japanese and Korean homes in the 1960s, life changed. Previously, the whole structure of the evening had been dictated by the need to produce steamed sticky white rice—the bedrock of every meal. The rice needed soaking, washing, and careful watching as it cooked in an earthenware pot, lest it burn.

The rice cooker—a bowl with a heating element underneath and a thermostat—removed all this work and worry. In today's versions, you just measure out the rinsed rice and water, and flip the switch. The thermostat tells the cooker when the water has been absorbed, and it switches from hot to warm. More deluxe cookers keep the rice warm for many hours and even have a time-delay function so that you can set the cooker before you leave for work.

Rice cookers were an ideal match between culture and technology. Early models replicated the slow simmering of a traditional earthenware Japanese rice pot. Unlike the microwave, which changed the entire structure of family meals, rice cookers enabled Asian families to eat the same traditional meals, but with far greater ease.

"Where There Are Asians, There Are Rice Cookers" is the title of a 2009 monograph by Yoshiko Nakano. Forget TVs, rice cookers are the most important electrical gadget in the Japanese home. Yet it's all happened remarkably fast. Electric rice cookers belong to the "Made in Japan" electronics boom of the 1950s. The first automated rice cooker was launched by Toshiba in 1956. In 1964, less than ten years later, the rate of rice-cooker ownership in Japan was 88 percent. From Japan, they traveled to Hong Kong, mainland China, and South Korea (where new cookers were designed with added pressure, to cook the rice softer, which is how Koreans like it). In tiny rural kitchens in China, the rice cooker may be the only stove, used to make gooey congee (rice porridge) as well as steamed rice.

What rice cookers are not so good for—thus far—are the long-grain rices of India and Pakistan. Basmati grains should be fluffy and separate. The slow steaming of the rice cooker does long grains no favors; they turn gummy. Which may explain why India does not yet fully share China's rice-cooker addiction.

KNIFE

The poet with his pen, the artist with his brush,
the cook with his chopping knife.

F. T. CHENG, 1954

I WAS MAKING A PILE OF CUCUMBER SANDWICHES ONE
day when I sliced off a sliver of finger instead of cucumber.
My injury was the result of getting overexcited with a Japanese mandolin slicer (newly acquired). "Lady with a mandolin," they shouted with cheery nonchalance, when I arrived at the
ER: clearly I was not the first idiot to hurt herself with this relatively
obscure gadget. Many enthusiastic cooks have a mandolin permanently discarded in some neglected cupboard, spattered with dried
blood. "Watch your fingers!" it said on the box, which should have
given me a clue, but somehow the thrill of seeing a heap of trans-
parent cucumber disks emerge distracted me, and before I knew it,
there was a slice of myself on the wrong side of the blade, lying
among the cucumber. It could have been worse. As I waited for the
paramedics, I felt a stab of relief that I had put the mandolin on its
thinnest setting.

Kitchens are places of violence. People get burned, scarred, frozen, and above all, cut. After the mandolin incident, I booked myself into a knife skills course, in a shiny new cooking school on the outskirts of town. Most of the men in the course had been given their enrollment as a present by wives and girlfriends, the assumption being that knives are the sort of thing men have fun with, like train sets or drills. They approached the chopping board with a slight swagger. The women stood more diffidently at first. We had without exception signed up for it ourselves, either as a treat (like yoga) or to get over some terror or anxiety around blades (like a self-defense class). I hoped it would teach me how to dice like a samurai, hack like a butcher, and annihilate an onion at ten paces like the chefs on TV. In fact, most of the course was about safety: how to hold vegetables in a clawlike grip with our thumbs tucked under, keeping knuckles always against the body of the knife so that we couldn't inadvertently baton our thumbs along with our carrots; how to steady the chopping board with a damp cloth; how to store our knives in a magnetic strip or in a plastic sheath. Our terror, it seemed, was justified. The teacher—a capable Swedish woman—warned us of the horrible accidents that ensue when sharp knives are carelessly left in a bowl of sudsy dishwashing detergent. You forget the knives are there, plunge your hand in, and slowly the water turns red, like a scene from *Jaws*.

Culinary knives have always been just a step away from weapons. These are tools designed to break, disfigure, and mutilate, even if all you are cutting is a carrot. Unlike lions, we lack the ability to tear meat from a carcass with our bare teeth; so we invented cutting tools to do the job for us. The knife is the oldest tool in the cook's armory, older than the management of fire by somewhere between 1 million and 2 million years, depending on which anthropologist you believe. Cutting with some implement or other is the most basic way of processing food. Knives do some of the work that feeble human teeth cannot. The earliest examples of stone cutting tools date back 2.6 million years to Ethiopia, where excavations have found both sharpened rocks and bones with cut-marks on them, indicating that raw meat was being hacked from the bone. Already, there was some sophistication in the knife skills on display. Stone Age humans fashioned numerous different cutting devices to suit their needs: archaeologists have identified simple sharp choppers, scrapers (both heavy duty and light duty), and hammerstones and spheroids for beating food. Even at this early stage, man was not randomly slashing at his food but making careful decisions about which cuts to make with which tools.

Unlike cooking, toolmaking is not an exclusively human activity. Chimpanzees and bonobos (another type of ape) have shown themselves capable of hammering rocks against other rocks to make sharp implements. Chimps can use stones to crack nuts and twigs to scoop fruit from a husk. Apes have also hammered stone flakes, but there is no evidence that they passed down toolmaking skills from one individual to another, as hominids did. Moreover, primates seem to be less sensitive to raw materials used than their human counterparts. Right from the beginning, hominid toolmakers were intensely interested in finding the best rocks for cutting, rather than just the most convenient, and were prepared to travel to get them. Which rock would make the sharpest flake? Stone Age toolmakers experimented with granite and quartz, obsidian and flint. Knife manufacturers still search for the best materials for a sharp blade; the difference is that the art of metallurgy, from the Bronze Age onward, has vastly expanded our options. From bronze to iron; from iron to steel; from steel to carbon steel, high-carbon steel, and stainless steel; and on to fancy titanium and laminates. You can now spend vast sums on a Japanese chef's knife, handmade by a master cutler from molybdenum-vanadium-enriched steel. Such a knife can perform feats that would have amazed Stone Age man, swooshing through a pumpkin's hard skin as if it were a soft pear.

In my experience, when you ask chefs what their favorite kitchen gadget is, nine times out of ten, they will say a knife. They say it slightly impatiently, because it's just so obvious: the foundation of every great meal is accurate cutting. A chef without a knife would be like a hairdresser without scissors. Knife work—more than the application of heat—is simply what chefs do: using a sharp edge to convert ingredients into something you can cook with. Different chefs go for different knives: a curved scimitar; a straight French "blooding" knife, designed for use by horsemeat butchers; a pointy German slicer; a cleaver. I met one chef who said he used a large serrated bread knife for absolutely everything. He liked the fact that he didn't have to sharpen it. Others favor tiny parers that dissect food with needle-sharp accuracy. Most rely on the classic chef's knife, either nine-inch or ten-inch, because it's about the right size to cover most needs: long enough for jointing, small enough for filleting. A good chef steels his or her knives several times in a single shift, drawing the blade swiftly to and fro at a 20° angle, to ensure the knife never loses its bite.

The story of knives and food is not only about cutting tools getting ever sharper and stronger, however. It is also about how we manage the alarming violence of these utensils. Our Stone Age ancestors took the materials at their disposal and—so far as we can surmise—made them as sharp as possible. But as the technology of knife making developed through iron and steel, the sharpest knife became something casually lethal. The primary function of a knife is to cut; but the secondary question has always been how to tame the knife's cutting power. The Chinese did it by confining their knife work to the kitchen, reducing food to bite-sized pieces with a massive cleaverlike instrument, out of sight. Europeans did it, first, by creating elaborate rules about the use of the knife at the table—the subtext of all table manners is the fear that the man next to you may pull his knife on you—and second, by inventing "table knives" so blunt and feeble that you would struggle to use them to cut people instead of food.

T here is a peculiar joy in holding a knife that feels just right for your hand and marveling as it dices an onion, almost without effort on your part. During the knife-skills course, our teacher showed us how to joint a chicken. When separating the legs from the thighs, you look for two little mountain tops; on hitting the right spot, the knife goes through like silk. This only works, however, when the knife is sharp enough to begin with.

Chefs always say that the safest knife is the sharpest one (which is true until you actually have an accident). Among domestic cooks, though, knowledge of how to keep a knife sharp has become a private passion rather than a universal skill. The travelling Victorian knife grinder, who could sharpen a set of knives in a matter of minutes—in exchange for whatever you could pay him, pennies or even a pint of ale—is long gone.* He has been replaced by eager knife enthusiasts, who grind their knives not as a job or even out of necessity but for the sheer satisfaction of it, swapping tips in online knife forums. Opinion differs as to which sharpening device is best for achieving the perfect edge, whether a Japanese water stone, a traditional whetstone, an Arkansas stone, or a synthetic aluminium oxide stone. Then there are the electric ones. Early models tended to be aggressive, but the latest generation, with spring-loaded slots for blades of varying dullness, are effective and precise.

Whichever tool is chosen, the basic principle is always the same. You sharpen a knife by grinding off some of the metal, starting with a coarse abrasive and moving to a finer one until you have the required sharpness. In addition, you may wish to steel your blade each time you use it, running it along a steel rod a few times

---

*If you look on the Internet, however, there are still a few knife-sharpening workshops that will sharpen anything from hunting knives to pizza wheels and food-processor blades.

to realign the edge. Steeling can keep a sharp knife sharp, but it cannot make a blunt knife sharp in the first place. What does it mean for a knife to be sharp? It is a question of angle. You get a sharp edge when two surfaces—known as bevels—come together at a thin V-shaped angle. If you could take a cross section through a sharp knife, you would see that the typical angle for a Western kitchen knife is around 20°, or one-eighteenth of a circle. European knives are generally double-beveled, that is, sharpened on both sides of the blade, resulting in a total angle of 40°. Every time you use your knife, a little of the edge wears away, and the angle is gradually lost. Sharpeners renew the edge by grinding bits of the metal away from both sides of the V to restore the original angle. With heavy use and excessive grinding, the blade gradually diminishes.

In an ideal universe, a knife would be able to achieve an angle of zero—representing infinite sharpness. But some concessions have to be made to reality. Thin-angled knives cut better—like razors—but if they are too thin, they will be too fragile to withstand the act of chopping, which rather defeats the purpose. Whereas Western kitchen knives sharpen at an angle of around 20°, Japanese ones, which are thinner, can sharpen at around 15°. This is one of the reasons so many chefs prefer Japanese knives.

There is much that the community of knife enthusiasts disagree upon. Is the best knife large? There's a theory that heavier knives do more of the work for you. Or small? There's another theory that heavy knives make your muscles ache. Are you better off with a flat edge or a curved one? The enthusiasts also disagree on the best way to test the sharpness of an edge to see if it "bites." Should you use your thumb—thus flaunting how at-one you are with the metal—or is it better to cut into a random vegetable or a ballpoint pen? There's a joke about a man who tested his blade using his tongue: sharp blades taste like metal; really sharp blades taste like blood.

What unites knife enthusiasts is the shared knowledge that having a sharp knife, and mastery of it, is the greatest power you will ever feel in a kitchen.

Shamefully late in my cooking life, I have discovered why most chefs think the knife is the one indispensable tool. You no longer feel anxious around onions or bagels. You look at food and see that you can cut it down to any size you want. Your cooking takes on a new refinement. An accurately chopped onion—tiny dice, with no errant larger chunks—lends a suave luxury to a risotto, because the onion and the grains of rice meld harmoniously. A sharp bread knife creates the possibility of elegantly thin toast. Become the boss of a sharp knife, and you are the boss of the whole kitchen.

This shouldn't really come as a revelation. But proficiency with a knife is now a minority enthusiasm. Even many otherwise accomplished cooks have a rack stocked with dull knives. I know, because I used to be one of them. You can survive perfectly well in the modern kitchen without any survivalist knife skills. When something needs to be really finely chopped or shredded, a food processor will pick up the slack. We are not in the Stone Age (as much as some of the knife enthusiasts would like us to be). Our food system enables us to feed ourselves even when we lack the most rudimentary cutting abilities, never mind the ability to make our own slicing tools. Bread comes pre-sliced, and vegetables can be bought pre-diced. Once, though, effective handling of a knife was a more basic and necessary skill than either reading or writing.

In medieval and Renaissance Europe, you carried your own knife everywhere with you and brought it out at mealtimes when you needed to. Almost everyone had a personal eating knife in a sheath dangling from a belt. The knife at a man's girdle could equally well be used for chopping food or defending himself against enemies. Your knife was as much a garment—like a wristwatch now—as a tool. A knife was a universal possession, often your most treasured one. Like a wizard's wand in *Harry Potter,* the knife was tailored to its owner. Knife handles were made of brass, ivory, rock crystal, glass, and shell; of amber, agate, mother of pearl, or tortoiseshell. They might be carved or engraved with images of babies, apostles, flowers, peasants, feathers, or doves. You would no more eat with another person's knife than you would brush your teeth today with a stranger's toothbrush. You wore your knife so habitually that—as with a watch—you might start to regard it as a part of yourself and forget it was there. A sixth-century text (St. Benedict's Rule) reminded monks to detach their knives from their belts before they went to bed, so they didn't cut themselves in the night.

There was a serious danger of this because knives then, with their daggerlike shape, really were sharp. They needed to be, because they might be called upon to tackle everything from rubbery cheese to a crusty loaf. Aside from clothes, a knife was the one possession every adult needed. It has been often assumed, wrongly, that knives, as violent objects, were exclusively masculine. But women wore them, too. A painting from 1640 by H. H. Kluber depicts a rich Swiss family, preparing to eat a meal of meat, bread, and apples. The daughters of the family have flowers in their hair, and dangling from their red dresses are silvery knives, attached to silken ropes tied around their waists. With a knife this close to your body at all times, you would have been very familiar with its construction.

Sharp knives have a certain anatomy. At the tip of the blade is the point, the spikiest part, good for skewering or piercing. You might use a knife's point to slash pastry, flick seeds from a lemon half, or spear a boiled potato to check if it is done. The main body of the blade—the lower cutting edge—is known as the belly, or the curve. This is the part of a knife that does most of the work, from shredding greens into a fine tangle to slicing meat thinly. Turn it on its side and you can use it to pulverize garlic to a paste with coarse salt: good-bye, garlic press! The opposite end of the blade from the belly, logically enough, is the spine, the blunt top edge that does no cutting but adds weight and balance. The thick, sharp part of the blade next to the handle is the heel, good for hefty chopping of hard things like nuts and cabbages. The blade then gives way to the tang, the piece of metal hidden inside the handle that joins the knife and its handle together. A tang may be partial—if it only extends partway into the handle—or it may be full. In many high-end Japanese knives now, there is no tang, the whole knife, handle and all, being formed from a single piece of steel. Where the handle meets the blade is called the return. At the bottom of the handle is the butt—the very end of the knife.

When you start to love knives, you come to appreciate everything from the quality of the rivets on the handle to the line of the heel. These are now fairly arcane thrills, but once, they belonged to everyone. A good knife was an object of pride. As you took it from your waist, the familiar handle, worn and polished from use, would ease nicely into your hand as you sliced your bread, speared your meat, or pared your apple. You knew the value of a sharp knife, because without it you would find it so much harder to eat much of what was on the table. And you knew that sharpness meant steel, which by the sixteenth century was already the metal most valued by knife makers.

The first metal knives were made of bronze during the Bronze Age (c. 3000–700 BC). They looked similar to modern knives in that as well as a cutting edge, they had a tang and bolster, onto which a handle could be fitted. But the cutting edge didn't function well because bronze is a terrible material for blades, too soft to hold a really sharp edge. Proof that bronze does not make good knives is confirmed by the fact that during the Bronze Age, cutting devices continued to be made from stone, which was, in many respects, superior to the newfangled metal.

Iron made better knife metal than bronze. The Iron Age was the first great knife age, when the flint blades that had been in use since the time of the Oldowans finally vanished. As a harder metal, iron could be honed far sharper than bronze. It was a handy metal for forging large, heavy tools. Iron Age smiths made pretty decent axes. For knives, however, iron was not ideal. Although harder than bronze, iron rusts easily, making food taste bad. And iron knives still do not hold the sharpest of edges.

The great step forward was steel, which is still, in one form or other, the material from which almost all sharp knives are made, the exception being the new ceramic knives, which have been described as the biggest innovation in blade material for three millennia. Ceramic knives cut like a dream through soft fish fillets or yielding tomatoes but are far too fragile for heavy chopping. For a blade that is sharp, hard enough, and strong, nothing has yet supplanted steel, which can form and hold a sharp edge better than other metals.

Steel is no more than iron with a tiny proportion of added carbon: around 0.2 percent to 2 percent by weight. But that tiny bit makes all the difference. The carbon in steel is what makes it hard enough to hold a really sharp angle, but not so hard it can't be sharpened. If too much carbon is added, the steel will be brittle and snap under pressure. For most food cutting, 0.75 percent carbon is right: this creates a "sheer steel" capable of being forged into chopping knives with a tough, sharp edge, easy to sharpen, without being easily breakable; the kind of knife that could cut almost anything.

By the eighteenth century, methods of making carbon steel had industrialized, and this marvelous substance was being used to make a range of increasingly specialized tools. The cutlery trade was no longer about making a daggerlike personal possession for a single individual. It was about making a range of knives for highly specific uses: filleting knives, paring knives, pastry knives, all from steel.

These specialized knives were both cause and consequence of European ways of dining. It has often been observed that the French haute cuisine that dominated wealthy European tastes from the eighteenth century on was a cuisine of sauces: béchamel, velouté, espagnole, allemande (the four mother sauces of Carême, later revised as the five mother sauces of Escoffier, who ditched the allemande and added hollandaise and tomato sauce). True, but it was no less a cuisine of specialist knives and precision cutting. The French were not the first to use particular knives for particular tasks. As with much of French cuisine, their multitude of knives can be traced to Italy in the sixteenth century. In 1570, Bartolomeo Scappi, Italian cook to the pope, had myriad kitchen weapons at his disposal: scimitars for dismembering, thick-bladed knives for battering, blunt-ended pasta knives, and cake knives, which were long, thin scrapers. Yet Scappi laid down no exact code about how to use these blades. "Then beat it with knives," Scappi will say, or "cut into slices." Again, he does not formally catalog different cutting techniques. It was the French, who, with their passion for Cartesian exactness, made knife work into a system, a rule book, and a religion. The cutlery firm Sabatier first produced carbon steel knives in the town of Thiers in the early 1800s—around the time that gastronomy as a concept was invented, through the writings of Grimod de la Reynière and Joseph Berchoux and the cooking of Carême. The knives and the cuisine went hand in hand. Wherever French chefs traveled, they brought with them a series of strict cuts—mince, chiffonade, julienne—and the knives with which to make them.

French food, no matter how simple, has meticulous knife work behind it. A platter of raw oysters on the half shell at a Parisian restaurant doesn't look like cooking at all, but what makes it a pleasure to eat, apart from freshness, is that someone has skillfully opened each mollusk with an oyster shucker, sliding the knife upward to cut the adductor muscle that holds the shell closed without smashing off any sharp pieces. As for the shallot vinegar with which the oysters are served, someone has had to work like crazy, cutting the shallot into brunoise: tiny 0.25 cm cubes. It is only this prepping that prevents the shallot from being too overwhelming against the bland, saline oysters.

The savory French steak that sits before you so invitingly—whether onglet, pavé, or entrecôte—is the result of French butchery using particular utensils: a massive cleaver for the most brutal bone-hacking work, a delicate butcher's knife for seaming out the more elusive cuts, and perhaps a cutlet basher (*batte à côtelettes*) to flatten the meat a little before it is cooked. The classical French kitchen includes ham knives and cheese knives, knives for julienning, and beak-shaped knives for dealing with chestnuts.

Professional haute cuisine was founded on specialism. The great chef Escoffier, who laid the foundation for all modern French restaurant cooking, organized the kitchen into separate stations for sauces, meats, pastries. Each of these units had its own persnickety knives. In a kitchen organized on Escoffier principles, one person might be given the job of "turning" potatoes into perfect little spheres. For this task, he would use a tournet knife, a small parer with a blade like a bird's beak. This curved blade would be awkward for cutting on a board—the angle is all wrong. Yet that arc is just right for swiping the skin off a handheld round object, following its contours to leave an aesthetically pleasing little globe. A garnish of turned vegetables—so pretty, so whimsical, so unmistakably French—is the direct result of a certain knife, wielded in a certain way, guided by a certain philosophy about what food should be.

Our food is shaped by knives. And our knives are fashioned by that mysterious combination of local resources, technological innovation, and cultural preferences that makes up a cuisine. The French way with knives is not the only way. In the case of China, an entire approach to eating and cooking was founded on a single knife, the *tou*, often referred to as the Chinese cleaver, perhaps the most fearsomely useful knife ever devised.

Cutting devices divide up into those that have one function and one function only—the Gorgonzola cutter, the arrow-shaped crab knife, the pineapple-slicing device that spirals down into the yellow fruit, removing the woody core and leaving only perfect juicy rings—and those that can be pressed into service for countless jobs: the multitaskers. And not surprisingly, different cooking cultures have produced different multitasker knives.

The Inuit *ulu*, for example, is a fan-shaped blade (similar to an Italian *mezzaluna*) traditionally used by Eskimo women for anything from trimming a child's hair to shaving blocks of ice, as well as chopping fish. The Japanese *santoku* is another multitasker, currently regarded as one of the most desirable all-purpose knives for the home kitchen. It is far lighter than a European chef's knife, with a rounded tip, and often has oval dimples, called divots, along the blade. *Santoku* means "three uses," so named because a *santoku* is equally good at cutting meat, chopping vegetables, and slicing fish.

Perhaps no knife is quite as multifunctional, nor quite as essential to an entire food culture, as the Chinese *tou*. This wondrous blade is often referred to as a "cleaver" because it has the same square-bladed hatchet shape as the cleaver that butchers use to hack through meat bones. The *tou's* use, however, is that of an all-purpose kitchen knife (for once, "all-purpose" is no exaggeration). For E. N. Anderson, the anthropologist of China, the *tou* exemplifies the principle of "minimax": maximum usage from minimum cost and effort. The idea is a frugal one: the best Chinese kitchen would extract the maximum cooking potential from the minimum number of utensils. The *tou* fits the bill. This big-bladed knife, writes Anderson, is useful for

> splitting firewood, gutting and scaling fish, slicing vegetables, mincing meat, crushing garlic (with the dull side of the blade), cutting one's nails, sharpening pencils, whittling new chopsticks, killing pigs, shaving (it is kept sharp enough, or supposedly is), and settling scores old and new with one's enemies

What makes the *tou* still more versatile is the fact that—unlike the Inuit *ulu*—it gave rise to what is widely considered one of the world's two greatest cuisines (the other being French). From ancient times, the great characteristic of Chinese cookery was the intermingling of flavors through fine chopping. The *tou* made this possible. During the Zhou dynasty (1045–256 BC), when iron was first introduced to China, the art of fine gastronomy was referred to as *"k'o'peng,"* namely, to "cut and cook." It was said of the philosopher Confucius (who lived from 551–479 BC) that he would eat no meat that had not been properly cut. By around 200 BC, cookbooks were using many different words for cutting and mincing, suggesting a high level of knife skills (*dao gong*).

A typical *tou* has a blade of around 18 to 28 cm (7 to 11 inches) long. So far, very similar to a European chef's knife. What's dramatically different is the width of the blade: around 10 cm, or 4 inches, nearly twice as wide as the widest point on a chef's knife. And the *tou* is the same width all the way along: no tapering, curving, or pointing. It's a sizable rectangle of steel, but also surprisingly thin and light when you pick it up, much lighter than a French cleaver. It commands you to use it in a different way from a chef's knife. Most European cutting uses a "locomotive" motion, rocking the knife back and forth, following the gradient of the blade. Because of its continuous flatness, a *tou* invites chopping with an up-down motion. The sound of knife work in a Chinese kitchen is louder and more percussive than in a French one: *chop-chop-chop* as opposed to *tap-tap-tap*. But this loudness does not reflect any crudeness of technique. With this single knife, Chinese cooks produce a far wider range of cutting shapes than the dicing, julienning, and so on produced by the many knives of French cuisine. A *tou* can create silken threads (8 cm long and very thin), silver-needle silken threads (even thinner), horse ears (3 cm slices cut on a steep angle), cubes, strips, and slices, to name but a few.

No single inventor set out to devise this exceptional knife, or if someone did, the name is lost. The *tou*—and the entire cuisine it made possible—was a product of circumstances. First, metal. Cast iron was discovered in China around 500 BC. It was cheaper to produce than bronze, which allowed for knives that were large hunks of metal with wooden handles. Above all, the *tou* was the product of a frugal peasant culture. A *tou* could reduce ingredients to small enough pieces that the flavors of all the ingredients in a dish melded together and the pieces would cook very quickly, probably over a portable brazier. It was a thrifty tool that could make the most of scarce fuel: cut everything small, cook it fast, waste nothing. As a piece of technology, it is much smarter than it first looks. In tandem with the wok, it works as a device for extracting the most flavor from the bare minimum of cooking energy. When highly chopped food is stir-fried, more of the surface area is exposed to the oil, becoming crispy-brown and delicious. As with all technology, there is a trade-off: the hard work and skill lavished on prepping the ingredients buys you lightning-fast cooking time. A whole, uncut chicken takes more than an hour to cook in the oven. Even a single chicken breast can take twenty minutes. But *tou*-chopped fragments of chicken can cook in five minutes or less; the time is in the chopping (though this, too, is speedy in the right hands; on YouTube you can watch chef Martin Yan breaking down a chicken in eighteen seconds). Chinese cuisine is extremely varied from region to region: the fiery heat of Sichuan; the black beans and seafood of the Cantonese. What unites Chinese cooks from distant areas is their knife skills and their attachment to this one knife.

The *tou* was at the heart of the way classical Chinese cooking was structured, and still is. Every meal must be balanced between *fan*—which normally means rice but can also apply to other grains or noodles—and *ts'ai*, the vegetables and meat dishes. The *tou* is a more essential component in this meal than any single ingredient, because it is the *tou* that cuts up the *ts'ai* and renders it in multiple different forms. There is an entire spectrum of cutting methods, with words to match. Take a carrot. Will you slice it vertically (*qie*) or horizontally (*pian*)? Or will you chop it (*kan*)? If so, what shape will you choose? Slivers (*si*), small cubes (*ding*), or chunks (*kuai*)? Whichever you adopt, you must stick to it exactly; a cook is judged by the precision of his or her knife strokes. There is a famous story about Lu Hsu, who was a prisoner under Emperor Ming. He was given a bowl of meat stew in his cell and knew at once that his mother had visited, for only she knew how to cut the meat in such perfect squares.

*Tou*s look terrifying. Handled by the right person, however, these threatening blades are delicate instruments and can achieve the same precision in cutting that a French chef needs an array of specialist blades to achieve. In skilled hands, a *tou* can cut ginger as thin as parchment; it can dice vegetables so fine they resemble flying-fish roe. This one knife can prepare an entire banquet, from cutting fragile slivers of scallop and 5 cm lengths of green bean to carving cucumbers to look like lotus flowers.

The *tou* is more than a device for fine dining. In poorer times, expensive ingredients can easily be omitted, so long as the knife work and the flavoring remain constant. The *tou* created a remarkable unity across the classes in Chinese cuisine, in contrast to British cookery, where rich food and poor food tend to operate in opposing spheres (the rich had roast beef, eaten from a tablecloth; the poor had bread and cheese, eaten from hand to mouth). Poor cooks in China might have far less *ts'ai*—far less vegetables and meat—to work with than their rich counterparts; but whatever they have, they will treat just the same. It is the technique, above all, that makes a meal Chinese or not. The Chinese cook takes fish and fowl, vegetable and meat, in all their diverse shapes and renders them geometrically exact and bite-sized.

The *tou*'s greatest power is to save those eating from any knife work. Table knives are viewed as unnecessary and also slightly disgusting in China. To cut food at the table is regarded as a form of butchery. Once the *tou* has done its work, all the eater has to do is pick up the perfectly uniform morsels using chopsticks. The *tou* and the chopsticks work in perfect symbiosis: one chops, the other serves. Again, this is a more frugal way of doing things than the classical French approach, where, despite all that laborious slicing with diverse knives in the kitchen, still further knives are needed to eat the meal.

The *tou* and its uses represent a radically different and alien culture of knives from that of Europe (and thence, America). Where a Chinese master cook used one knife, his French equivalent used many, with widely differing functions: butcher's knives and boning knives, fruit knives and fish knives. Nor was it just a question of implements. The *tou* stood for a whole way of life of cooking and eating, one completely removed from the courtly dining of Europe. There is a vast chasm between a dish of tiny dry-fried slivers of beef, celery, and ginger, done in the Sichuan style, seasoned with chili-bean paste and Shaoxing wine in a careful balance of flavors; and a French steak, bloodied and whole, supplied at the table with a sharp knife for cutting and mustard to add flavor, according to the whim of the diner. The two represent diverse worldviews. It is the gulf between a culture of chopping and one of carving.

In Europe, the pinnacle of knife work was not that performed by the cook but by the courtly carver, whose job it was to divide up meat at the dinner table for the lords and ladies. Whereas the *tou* was used on raw food and rendered it all as similar as possible, the medieval carver dealt with cooked food and was expected to understand that every animal—roasted whole—needed to be carved in its own special way with its own special knife and served with its own special sauce.

"Sir, pray teach me how to carve, handle a knife and cut up birds, fish and flesh," pleads one medieval courtesy book. According to a book published by Wynkyn de Worde in 1508, the English "Terms of a Carver" went thus:

> Break that deer
> Slice that brawn
> Rear that goose
> Lift that swan
> . . . Dismember that heron

The rules of carving belonged to a world of symbols and signs: each animal had its own logic and had to be divided up accordingly. There was a connection between the knives of carving and the weapons of hunting: the point was to divide the spoils of the hunt in a strict hierarchy to emphasize the power of the man on whose land the animals had been killed. The carver's knife had to follow the lines and sinews of any given beast, and to do so in the service of a lord; it could not strike freely like a *tou*. The carver had to know that the wings on a hen were minced, whereas the legs were left whole. Further, there was honor to be had in getting it right. Carving was seen as so important at court that it evolved into a special office, "the Carvership," which was held by designated officials and even included members of the nobility.

Unlike modern carvers, whose task is the equitable distribution of food as they preside over a Sunday roast or a Thanksgiving turkey, the medieval European carver was not in charge of the whole table but served only a single lord. His task was not sharing food fairly but rather taking the best of what was on the table for his particular master. He would scoop up samples of all the sauces on little pieces of bread, popping them into the mouths of waiters, to check for poison. A big part of the job was preventing the lord from consuming any "fumosities"—in other words, gristle, skin, feathers, or anything else that might prove indigestible. Beyond that, the carver didn't actually do all that much with his knife. The lord would have his own sharp knife, after all, with which to tackle the meat as he ate it.

What is striking about the medieval carving knife is how few cuts it made. The language of carving was brutal: *dismember, spoil, break, unjoint*. In contrast to the Chinese chef with his single *tou*, the knives at the carver's disposal were many: large, heavy knives for carving big roasts such as stags and oxen; tiny knives for game birds; broad spatula-like serving knives for lifting the meat onto the trencher; and thin, blunt-bladed credence knives for clearing all crumbs from the tablecloth. Yet very few knife strokes were actually performed on the roast meats. To "dismember a heron" is a chilling phrase, but what it actually involved was posing the poor dead bird in a supposedly elegant arrangement on the trencher rather than chopping it into tiny pieces: "Take a heron, and raise his legs and wings as a crane, and sauce him," says Worde. Sometimes the carver needed to break up large bones, and sometimes he would shred a bit of the meat—a capon wing was minced and mixed with wine or ale, for example. But the job of carver was more about serving than cutting. The carving knife did not need to render all of the food into bite-sized pieces. This would have been to usurp the role of the lord's own knife.

The habit of carrying your own sharp knife with you was as much a bedrock of Western culture as Christianity, the Latin alphabet, and the rule of law. Until, suddenly, it wasn't. So much of what we believe about utensils is determined by culture, but cultural values are not fixed and eternal. From the seventeenth century onward, there was a great upheaval in European attitudes toward knives. The first change was that knives started to be pre-laid on the table, joined by that newfangled implement, the fork. This divested knives of their former magic. Rather than being specially tailored to an individual owner, cases of identical knives were now bought and sold by the dozen and laid out impersonally for whomever happened to sit down. The second change was that table knives ceased being sharp. They were thus divested of their power, too. The raison d'être of knives is to cut. It takes a civilization in an advanced state of politesse—or passive aggression—to devise *on purpose* a knife that does a worse job of cutting. In more ways than one, we are still living with the consequences of this change today.

I n 1637, Cardinal Richelieu, chief adviser to King Louis XIII of France, is supposed to have witnessed a dinner guest using the sharp tip of a double-edged knife to pick his teeth. This act so appalled the cardinal—whether because of the danger or the vulgarity is not entirely clear—that he ordered all his own knives to be made blunt, starting a new fashion. Until that time, eating knives tended to be sharpened on both sides of the blade, like a dagger. No more. In 1669, cutlers were forbidden by the next king, Louis XIV, from forging pointed dinner knives in France. Richelieu's mandate against double-edged knives went along with a transformation of table manners and table implements. Europe underwent what the great sociologist Norbert Elias called the "Civilising process." Patterns of behavior at the dining table changed markedly. Old certainties were crumbling. The Catholic Church had lost its former unity, and the chivalric codes of behavior were long gone. People suddenly felt revolted by ways of eating that had once been acceptable: taking meat from a common dish using fingers, drinking soups straight from the bowl, and using a single sharp knife to cut everything. All these things—which were once entirely in keeping with courtly manners—now felt uncivilized. Europeans now shared the Chinese wariness of sharp knives at the table. Unlike the Chinese, Westerners kept knives for eating but disabled them in various ways.

In France, knives were often kept off the table, except for certain specific tasks such as peeling and cutting fruit, for which personal sharp knives were produced, as in the old days. English knives stayed on the table but became significantly blunter. Sixteenth- and seventeenth-century English table knives look like miniature kitchen knives. The shape of the blade may vary, from daggerlike to penknife straight to scimitar-bladed. Sometimes the blade is double-edged, sometimes single-edged. But the knives all have this in common: they are sharp (or at least they would have been when they were shiny and new).

Eighteenth-century table knives look completely different from those of the previous century. Suddenly, they are ostentatiously blunt. The blade often curves gently toward the right, finishing in a thoroughly rounded tip. It is a shape we now associate with butter knives—and with good reason. The table knife had ceased to be a very effective cutting device. It was now an ineffectual utensil, only good for spreading butter, placing things on the fork, or subdividing food that was already relatively soft.

The new toothless table knife also led to a change in the way knives were held. Previously, a knife might be grasped with the whole hand in a stabbing pose. Now, the index finger was poised delicately along the top of the—newly blunt—spine with the palm of the hand wrapped round the handle. This is still the polite way to hold a table knife. It is one of the reasons so many of us have bad knife skills. We use the same grip on sharp knives as table knives, which is disastrous. When holding a kitchen knife, you should never rest your index finger along the spine—there's far more danger of cutting yourself than when you robustly grip the bottom of the blade with thumb on one side and forefinger on the other. A good training in table manners—which teach constant diffidence around sharpness—is bad training for the kitchen.

By the eighteenth century, polite Westerners sat at the dinner table delicately holding their pretty little knives, trying to avoid at all cost any gesture reminiscent of violence or menace. As a cutting technology, the table knife was now more or less redundant. By the late eighteenth century, the celebrated Sheffield table knife from the north of England, though still made of top-notch steel, had become less about cutting and more about display. In London society, these were beautiful objects, laid out on the table as marks of a host's good taste and wealth. It would be easy to write off table knives as technologically obsolete in the modern era. The uselessness of table knives is shown by the appearance of sharp, serrated steak knives (pioneered in the southern French town of Laguiole), whose presence acts as a kind of rebuke to normal knives: what steak knives say is that when you actually need to cut something at the table, a table knife won't do.

The table knife was now an entirely separate object from the knife as weapon. There was no need to carry a knife with you; indeed, to do so could be considered poor form, in Britain at any rate. In 1769, an Italian man of letters, Joseph Baretti, was indicted for stabbing a man in self-defense in London, using a small folding fruit knife. Baretti's defense was that it was common practice on the Continent to carry a sharp knife for cutting apples, pears, and sweetmeats. The fact that he had to explain this in such detail to a British court was a sign of how the nature of knives had changed in Britain by 1769. Sharpness was no longer seen as necessary or even desirable in a table knife. In this, Britain was leading the way.

There is more to table knives than sharpness, however. There is also the question of how they make food more pleasurable to eat—or not. From this perspective, for most people, table knives really only came into their own in the twentieth century, with the advent of stainless steel.

I said earlier that the carbon steel favored by Sheffield cutlers was a far better metal for forging blades than previous alternatives. What I didn't mention was this: the downside of carbon steel, like iron, is that it can make certain foods taste disgusting. Anything acidic has a potentially disastrous effect on non–stainless steel. "Upon the slightest touch of vinegar," wrote the great American etiquette expert Emily Post, steel-bladed knives turned "black as ink." Vinaigrette and steel knives were a particularly bad combination, hence the French prejudice, that persists to this day, against cutting salad leaves.

Another problem was fish. For centuries, people have found lemon to be the ideal accompaniment to fish. But until the 1920s, and the invention of stainless steel, the taste of lemony fish was liable to be ruined by the tang of blade metal from the knife. The acid in the lemon reacted with the steel, leaving a foul metallic aftertaste that entirely overpowered the delicate flesh of the fish. This explains the production of fish knives made of silver in the nineteenth century. Nowadays, these seem a pointless affectation. In fact, fish knives were a mainly practical invention, albeit one that only the rich could afford. Unlike normal steel knives, silver knives were noncorrosive and did not react with the lemon juice on the plate. The signature scalloped shape was firstly a way to distinguish them in the cutlery drawer (as well as signaling the fact that fish, unlike meat, was not tough and did not need to be sawed at). If you had no silver fish knives, the only other option was to eat fish with two forks, or a single fork and a piece of bread, or suffer the taste of corroded steel.

So, the launch of stainless steel in the twentieth century ranks as one of the greatest additions to happiness at the table. Once it entered cheap mass production after World War II, it placed stylish, shiny cutlery within the reach of most budgets and removed at a stroke all those fears about knives making food taste funny. Never again would you have to worry when you squirted a lemon over a piece of cod or feel that you mustn't use a knife to cut dressed salad.

Stainless steel (otherwise known as inox steel or nonrusting steel) is a metal alloy with a high chromium content. The chromium in the metal forms an invisible layer of chromium oxide when exposed to the air, which is what enables stainless steel to remain resistant to corrosion and also splendidly lustrous. It was only in the early years of the twentieth century that a successful stainless steel—strong and tensile enough as well as corrosion resistant—was made. In 1908, Friedrich Krupp built a 366-ton yacht—*Germania*—with a chrome-steel hull. Meanwhile, in Sheffield, Harry Brearley of Thomas Firth and Sons had discovered a stainless steel alloy while trying to find a corrosion-resistant metal for making gun barrels. Noncorrosive cutlery was a happy by-product of the search for military advantage between Britain and Germany on the road to total war. At first, the new metal was hard to work in all but the simplest cutlery patterns; it took the industrial innovations of World War II for stainless steel knives to become something that could be worked efficiently and cheaply in the shapes people desired. Stainless steel was another step in domesticating the knife, in rendering it cheaper, more accessible, and less threatening than the knives our ancestors carried around on their person.

The Western table knife now seems an altogether harmless object (though they were still thought menacing enough to have been banned from planes in the wake of 9/11). Our preference for these blunt implements over the past two hundred years has had powerful unseen consequences, however.

Knives do not just leave their mark on food. They leave it on the human body. Every chef has scars to show, and often does so proudly, giving you the story behind each wound. Hack marks on a thumb from paring vegetables; the missing chunk of finger from an unfortunate encounter with a turbot. My finger still bulges tenderly where the mandolin sliced it. Then there are the blisters and calluses that chefs acquire, which appear without any accidents or mistakes, just through the action of good knife work. Blisters and gashes are the most obvious legacy of the kitchen knife, but the marks the knife has left on our bodies go further still. The basic technology of cutting food at the table has shaped our very physiology, and above all, our teeth.

Much of the science of modern orthodontics is devoted to creating—through rubber bands, wires, and braces—the perfect "overbite." An overbite refers to the way our top layer of incisors hangs over the bottom layer, like a lid on a box. This is the ideal human occlusion. The opposite of an overbite is the "edge-to-edge" bite seen in primates such as chimpanzees, where the top incisors clash against the bottom ones, like a guillotine blade.

What the orthodontists don't tell you is that the overbite is a very recent aspect of human anatomy and probably results from the way we use our table knives. Based on surviving skeletons, this has only been the "normal" alignment of the human jaw for 200 to 250 years in the Western world. Before that, most human beings had an edge-to-edge bite, comparable to apes. The overbite is not a product of evolution—the time frame is far too short. Rather, it seems likely to be a response to the way we cut our food during our formative years. The person who worked this out is Professor Charles Loring Brace (born 1930), a remarkable American anthropologist whose main intellectual passion was Neanderthal man. Over decades, Brace built up the world's largest database on the evolution of hominid teeth. He possibly held more ancient human jaws in his hand than anyone else in the twentieth century.

As early as the 1960s, Brace had been aware that the overbite needed explaining. Initially, he assumed that it went back to the "adoption of agriculture six or seven thousand years ago." Intuitively, it would make sense if the overbite corresponded to the adoption of grain, because cereal potentially requires a lot less chewing than the grainy meat and fibrous tubers and roots of earlier times. But as his tooth database grew, Brace found that the edge-to-edge bite persisted much longer than anyone had previously assumed. In Western Europe, Brace found, the change to the overbite occurred only in the late eighteenth century, starting with "high status individuals."

Why? There was no drastic alteration in the nutritional components of a high-status diet at this time. The rich continued to eat large amounts of protein-rich meat and fish, copious pastries, modest quantities of vegetables, and about the same amount of bread as the poor. Admittedly, the rich in 1800 would expect their meat to come with different seasonings and sauces than in 1500: fewer currants, spices, and sugar, but more butter, herbs, and lemon. Cooking styles certainly evolved. But most of these changes in cuisine long predated the emergence of the overbite. The fresher, lighter nouvelle cuisine that appeared on tables across Europe during the Renaissance goes back at least as far as 1651, with the French cookbook by La Varenne called *Le Cuisinier françois*; arguably, it goes back still further, to the Italian chef Maestro Martino in the 1460s, whose recipes included herb frittata, venison pie, parmesan custard, and fried sole with orange juice and parsley, all things that would not have looked out of place at wealthy dinners three hundred years later. At the time that aristocratic teeth started to change, the substance of a high-class diet had not radically altered in several hundred years.

What changed most substantially by the late eighteenth century was not *what* was eaten but *how* it was eaten. This marked the time when it became normal in upper- and middle-class circles to eat with a table knife and fork, cutting food into little pieces before it was eaten. This might seem a question of custom rather than of technological change, and to some extent it was. After all, the mechanics of the knife itself were hardly new. Over millennia, people have devised countless artificial cutting implements to make our food easier for our teeth to manage. We have hacked, sawed, carved, minced, tenderized, diced, julienned. The Stone Age mastery of cutting tools seems to have been one of the factors leading to the smaller jaws and teeth of modern man, as compared with our hominid ancestors. But it was only 200 to 250 years ago, with the adoption of the knife and fork at the dining table, that the overbite emerged.

In premodern times, Brace surmises that the main method of eating would have been something he has christened "stuff-and-cut." As the name suggests, it is not the most elegant way to dine. It goes something like this. First, grasp the food in one of your hands. Then clamp the end of it forcefully between your teeth. Finally, separate the main hunk of food from the piece in your mouth, either with a decisive tug of your hand or by using a cutting implement if you have one at hand, in which case you must be careful not to slice your own lips. This was how our ancestors, armed only with a sharpened flint, or, later, a knife, dealt with chewy food, especially meat. The "stuff-and-cut" school of etiquette continued long after ancient times. Knives changed—from iron to steel, from wood-handled to porcelain-handled—but the method remained.

The growing adoption of knife-and-fork eating in the late eighteenth century marked the demise of "stuff-and-cut" in the West. We will return to the fork (and the chopstick and the spoon) in Chapter 6. For the moment, all we need to consider is this. From medieval to modern times, the fork went from being a weird thing, a pretentious object of ridicule, to being an indispensable part of civilized dining. Instead of stuffing and cutting, people now ate food by pinning it down with the fork and sawing off little pieces with the table knife, popping pieces into the mouth so small that they hardly required chewing. As knives became blunter, so the morsels generally needed to be softer, reducing the need to chew still further.

Brace's data suggest that this revolution in table manners had an immediate impact on teeth. He has argued that the incisors—from the Latin *incidere,* "to cut"—are misnamed. Their real purpose is not to cut but to clamp food in the mouth—as in the "stuff-and-cut" method of eating. "It is my suspicion," he wrote, "that if the incisors are used in such a manner several times a day from the time that they first begin to erupt, they will become positioned so that they normally occlude edge to edge." Once people start cutting their food up very small using a knife and fork, and popping the morsels into their mouths, the clamping function of the incisors ceases, and the incisors continue to erupt until the top layer no longer meets the bottom layer: creating an overbite.

We generally think that our bodies are fundamental and unchanging, whereas such things as table manners are superficial: we might change our manners from time to time, but we can't be changed by them. Brace turned this on its head. Our supposedly normal and natural overbite—this seemingly basic aspect of modern human anatomy—is actually a product of how we behave at the table.

How can we be sure, as Brace is, that it was cutlery that brought about this change in our teeth? The short answer is that we can't. Brace's discovery raises as many questions as it answers. Modes of eating were far more varied than his theory makes room for. Stuff-and-cut was not the only way people ate in preindustrial Europe, and not all food required the incisor's clamp; people also supped soups and potages, nibbled on crumbly pies, spooned up porridge and polenta. Why did these soft foods not change our bite much sooner? Brace's love of Neanderthals may have blinded him to the extent to which table manners, even before the knife and fork, frowned upon gluttonous stuffing. Posidonius, a Greek historian (born c. 135 BC) complained that the Celts were so rude, they "clutch whole joints and bite," suggesting that polite Greeks did not. Moreover, just because the overbite occurs at the same time as the knife and fork does not mean that one was caused by the other. Correlation is not cause.

Yet Brace's hypothesis does seem the best fit with the available data. When he wrote his original 1977 article on the overbite, Brace himself was forced to admit that the evidence he had so far marshaled was "unsystematic and anecdotal." He would spend the next three decades hunting out more samples to improve the evidence base.

For years, Brace was tantalized by the thought that if his thesis was correct, Americans should have retained the edge-to-edge bite for longer than Europeans, because it took several decades longer for knife-and-fork eating to become accepted in America. After years of fruitless searching for dental samples, Brace managed to excavate an unmarked nineteenth-century cemetery in Rochester, New York, housing bodies from the insane asylum, workhouse, and prison. To Brace's great satisfaction, he found that out of fifteen bodies whose teeth and jaws were intact, ten—two-thirds of the sample—had an edge-to-edge bite.

What about China, though? "Stuff-and-cut" is entirely alien to the Chinese way of eating: cutting with a *tou* and eating with chopsticks. The highly chopped style of Chinese food and the corresponding use of chopsticks had become commonplace around nine hundred years before the knife and fork were in normal use in Europe, by the time of the Song dynasty (960–1279 AD), starting with the aristocracy and gradually spreading to the rest of the population. If Brace was correct, then the combination of *tou* and chopsticks should have left its mark on Chinese teeth much earlier than the European table knife.

The supporting evidence took a while to show up. On his eternal quest for more samples of teeth, Brace found himself in the Shanghai Natural History Museum. There, he saw the pickled remains of a graduate student from the Song dynasty era, exactly the time when chopsticks became the normal method of transporting food from plate to mouth.

> This fellow was an aristocratic young man, an official, who died, as the label explained, around the time he would have sat for the imperial examinations. Well, there he was, in a vat floating in a pickling fluid with his mouth wide open and looking positively revolting. But there it was: the deep overbite of the modern Chinese!

Over subsequent years, Brace has analyzed many Chinese teeth and found that—with the exception of peasants, who often retain an edge-to-edge bite well into the twentieth century—the overbite does indeed emerge 800–1,000 years sooner in China than in Europe. The differing attitude to knives in East and West had a graphic impact on the alignment of our jaws.

The knife as a technology goes beyond sharpness. The way a knife is used matters just as much as how well it slices. The *tou* that cut this Chinese aristocrat's food a thousand years ago would not have been significantly sharper or stronger than the carving knives that were cutting the meat of his European counterparts at the time. The greatest difference was what was done with it: cutting raw food into tiny fragments instead of carving cooked food into large pieces. The cause of this difference was cultural, founded on a convention about what implements to use at the table. Its consequences, however, were starkly physical. The *tou* had left its mark on the Chinese student's teeth, and it was there for Brace to see.

～ Mezzaluna ～

WITH ITS STUBBY WOODEN HANDLES AND
arching blade, the *mezzaluna* looks like an implement that should have fallen out of use several
centuries ago. Some version of this curved mincing
knife has been in kitchens at least since the Renaissance in Italy. Before the *mezzaluna*, Italian
cooks employed many single-handled curved
knives. There were also double-handled knives,
but they were for scraping the table clean, not
chopping. Finally, some enterprising palace smith
must have thought to combine the sharp curved
blade with the double handle, to create the perfect
utensil for mincing. And still the *mezzaluna* en-
dures, chopping herbs and lending its pretty
name—"half-moon" in Italian—to numerous upscale restaurants.

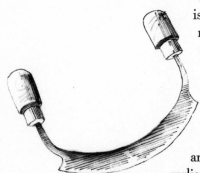

The *mezzaluna*'s staying power
is a warning not to underesti-
mate the power of romance in
the kitchen. This is a thrilling
object to use. It's like taking
your hands for a swing-boat
ride in some ancient Italian
city. Up-down, up-down. You
look down and inhale the giddy
aroma of parsley, lemon peel, and
garlic—the gremolata you've made to
sprinkle on an osso bucco.

Yes, you could have blitzed it in a food processor or chopped it with a regular chef's knife. But
the *mezzaluna* does it better. There is efficiency

behind the romance. When chopping nuts, for example, processors have a tendency to overdo things—you hold the pulse button too long and before you know it, you have ground almonds; a minute later, nut butter. With a chef's knife, the nuts skitter all over the board. The *mezzaluna* catches the nuts at each end as it rocks, producing a nicely uneven rubble in no time.

Single-bladed *mezzalunas* are best because with a double blade, what you gain in power, you waste in time as you push out debris that clumps between the blades. A single curved blade is easily powerful enough to dispatch dried apricots, which gum up normal knives. And its rocking motion remains the best way to chop fresh green herbs until they are fine but not mush.

The *mezzaluna* has another great advantage over the knife, which cooking guru Nigella Lawson points out. With the *mezzaluna*, she writes, "both my hands are engaged and thus it is impossible for me to cut myself."

❧ 3 ❧

FIRE

Probably the greatest [discovery],
excepting language, ever made by man.
    CHARLES DARWIN, on cooking

O father, the pig, the pig, do come
and taste how nice the burnt pig eats.
    CHARLES LAMB, "A Dissertation upon Roast Pig"

I MAGINE DOING THIS IN AN UNLIT KITCHEN—SEE HOW
dangerous it is!" A man dressed in a black T-shirt and white
chef's apron is standing near a hot fire, thrusting a small piece
of veal stuffed with sage leaves onto what looks like an instrument of torture. It is composed of five deadly iron spears, each several
feet long and precariously joined together. This device looks like a
five-pronged javelin. It is actually a rare type of spit called a *spiedo
doppio*, an Italian device for roasting meats from the sixteenth century.
The man holding it is Ivan Day. He
may be the only person in the
world who still cooks with one.

Day, a boyish man in his early sixties, is the foremost historian of food in Britain. He lives in the Lake District in a rickety seventeenth-century farmhouse, crammed with period utensils and antiquarian cookbooks, a kind of living museum where he gives courses on historic cookery. Day teaches groups of amateur cooks (as well as numerous chefs, scholars, and museum curators) how to cook historically. In an Ivan Day course, you might learn how to make a Renaissance pie of quinces and marrow bone; a seventeenth-century wafer flavored with rosewater; Victorian jelly; or medieval gingerbread, all made with the authentic equipment. Day's greatest passion, though, is for spit-roasting, which he believes to be the finest technique ever devised for cooking meat. "People tell me my roast beef is the best they have ever tasted," he observes in one of his courses. His hearth and all its spits enable him to roast vast joints, sometimes seventeen pounds at a time.

Standing on the uneven stone floor in Ivan Day's kitchen, I am struck by how unusual it now is to have an entire house organized around an open hearth. Once, almost everyone lived like this, because a single fire served to warm a house, heat water for washing, and cook dinner. For millennia, all cooking was roasting in one form or another. In the developing world, the heat of an open fire remains the way that the very poorest cook.

But in our own world, fire has been progressively closed off. Only at barbecues or campfires, sitting around toasting marshmallows and warming our hands by the flames, do we encounter a cooking fire directly. Many of us proclaim a fondness for roast beef—and Ivan Day's really was the best I've ever tasted—but we have neither the resources nor the desire to set up our homes in the service of open-hearth cookery. We have plenty of other things to do, and our cooking has to fit around our lives, rather than the other way around. It takes huge effort on Day's part to maintain this kitchen. Day scours the antiquarian markets of Europe looking for spits and other roasting utensils, all of which got junked many decades ago when kitchens were converted away from open hearths to closed-off stoves and cooktops.

It is not just a question of the fire itself. Cooking by an open hearth went along with a host of related tools: andirons or brand-irons to stop logs rolling forward at either end of the fire; hasteners, which are large metal hoods placed in front of the fire to speed up the cooking or protect the cook from the heat; spits of numerous kinds, from small and single-pronged to vast and five-pronged; spit-jacks to rotate the meat on the spit; fire tongs and bellows to control the fire; pot hooks for hanging pots over the fire and dripping pans to go under the fire to catch the fat dripping off roasting meat; brandreths and trivets to support cooking pots, and flesh-forks for pulling pieces of meat out of the pot. All these implements were made of heavy metal (usu-

ally iron) and were long-handled to protect the cook from the fierce heat. Not one of these things can be found in kitchenware shops today. They vanished along with the open hearth.

If I came into Day's kitchen with short-handled stainless steel tongs and nonstick silicone spatulas, I wouldn't stand a chance. The utensils would melt. I would fry. The children would howl. Dinner would burn. The entire way of life that supported cooking by an open hearth has become obsolete. Kitchen technology is not just about how well something works on its own terms—whether it produces the most delicious food—but about all the things that surround it: kitchen design; our attitude to danger and risk; pollution; the lives of women and servants; how we feel about red meat, indeed about meat in general; social and family structures; the state of metallurgy. Roasting meat before the fire goes along with an entire culture that has been lost. This is why it is so disconcerting to step into the kitchen of

Ivan Day, one of the last men in Britain who is prepared to build his life around an open fire.

Roasting is the oldest form of cooking. At its most basic, it means nothing more than placing raw ingredients directly into a fire. In Africa, the Kung!San hunter-gatherers still cook like this, plunging tsin beans into hot ash. We will never know the lucky person who—whether by accident or design—first discovered that food could be transformed by fire, becoming both easier to digest and more delicious. In his "A Dissertation upon Roast Pig," Charles Lamb imagines it taking place in China when Bo-bo, the lazy son of a swineherd, starts a house fire that kills, and accidentally burns, a litter of piglets. In Lamb's fable, Bo-bo marvels at the savory smell. He reaches out to take a fragment of hot scorched pig skin, "and for the first time in his life (in the world's life, indeed, for before him no man had known it) he tasted—crackling!"

It's an alluring story, but the discovery of roasting can't possibly have happened like this, for the obvious reason that roast meats long predate both houses and swineherds. The technology of roasting is far older than that of constructing buildings, and older still than agriculture. It predates both pottery for boiling and ovens for baking by nearly 2 million years. The oldest building known dates to around half a million years ago, toward the end of the era of *Homo erectus,* the first hunter-gatherer humans. It would be many thousands of years, however, before these house-dwelling proto-humans became farmers. Plant agriculture dates back to around 10,000 BC, well into the time of modern man or *Homo sapiens*. Animal husbandry is yet more recent. Pigs were only domesticated in China around 8000 BC. By this time, our ancestors had already been familiar with the savory taste of roast meat for hundreds of thousands of years.

Indeed, it may have been the discovery of roasting over an open fire that first made us what we are. If anthropologist Richard Wrangham is correct, this first act of cooking or roasting—around 1.8 to 1.9 million years ago—was *the* decisive moment in history: namely, the moment when we ceased to be upright apes and became more fully human. Cooking makes most foods far easier to digest, as well as releasing more of the nutritive value. The discovery of cooked food left us with surplus energy for brain growth. Wrangham writes that "cooking was a great discovery not merely because it gave us better food, or even because it made us physically human. It did something even more important: it helped make our brains uniquely large, providing a dull human body with a brilliant human mind."

Having tamed this potent source of heat and light, men built homes near to it, and then around it. The hearth that supplied every meal was always the focal point of the house. Indeed, the Latin word *focus* translates as "fireplace." The need to maintain a fire—to start it, to keep it going at the right heat, to supply it with enough fuel during the day and to damp it down at night so that the house didn't burn down—these were the dominant domestic activities until 150 years ago, with the coming of gas ovens. The term *curfew* now means a time by which someone—usually a teenager—has to get home. The original curfew was a kitchen object: a large metal cover placed over the embers at night to contain the fire while people slept. As for cooking itself, it was largely the art of fire management.

In the modern kitchen, fire has not just been tamed. It has been so boxed off, you could forget it existed at all, amid the cool worktops and all the on-off switches that enable us to summon heat and dismiss it again in a second. But then fire resurfaces and reminds us that even in the modern world, kitchens are still places where people get burned. In a Greek study of 239 childhood burn cases, it was found that the kitchen was by far the most dangerous room of the house, causing 65 percent of burn injuries. The age group most affected by kitchen scalds are one-year-olds: old enough to be mobile, not old enough to know that stoves are hot.

In earlier times, you walked into a kitchen and expected to see fire. Now, the presence of fire is a signal to panic. In the United Kingdom today, the majority of fires in the home are still caused by cooking, specifically by leaving pans of food unattended, and even

more specifically by leaving chip pans unattended. The chip pan—a
deep-sided open pan in which potatoes are fried in a basket—is an
interesting example of how people often cling to kitchen technologies long after they have been proven to be lethal and inefficient.
There are around 12,000 chip pan fires in the UK every year, resulting in 4,600 injuries and fifty deaths. The fire services periodically
plead with the public to give up cooking fries in chip pans, begging
people either to buy a proper deep-fat fryer with a closed lid or just
eat something—anything!—else instead, particularly when drunk.
But still the chip-pan fires continue.

The great British chip-pan fire is emblematic of a deep forgetful-
ness, which goes beyond the obvious dopiness of combining drinking with hot oil late at night in confined spaces. There is a sort of
innocence about the chip-pan blaze, as if those responsible had altogether forgotten the connection between cooking and fire. Real,
deadly fire. This was not something you could ever forget in the days
when all cooking started with an open flame.

Brillat-Savarin, the great French philosopher of cuisine, wrote in
1825 that "a cook may be taught but a man who can roast is born
with the faculty." The first time I read this, as a student getting
started in the kitchen, I was puzzled. Roasting didn't seem that hard
to me—certainly not compared with making mayonnaise that didn't
separate or a puff pastry that didn't fall apart. It was no trouble to
dab a three-pound chicken with butter, salt, and lemon, put it in
roasting dish in a hot electric oven, wait an hour and ten minutes,
then remove. So long as I bought a good free-range bird, my "roast
chicken" came out perfect every time. Roasting was far easier than
braising a shin of beef or sautéing a pork chop, both of which required stringent attention to ensure the meat didn't toughen.

This basic procedure was not at all what Brillat-Savarin had in
mind. Until well into the nineteenth century, there was a strict con-
ceptual division in Western cookery between open fires—things that
roasted; and closed ovens—things that baked. To Brillat-Savarin,

what I do with a chicken has little to do with roasting. From the point of view of most cooks of previous centuries, the "roast dinners" we serve up are nothing of the kind but are instead a strange kind of baked meats, half broiled, half stewed in their own fat. The point about roasting in its original sense was that it required, firstly, an open hearth and, secondly, rotation on a spit (the root of the word *roast* is the same as "rotate").

The original direct-fire roasting—shoving something into an untamed fire—is a crude and quick method that results in chewy, greasy meat. The muscle protein gets overcooked and chewy, while collagen in the connective tissue does not have time to tenderize. True roasting, by contrast, is a gentle process. The food cooks at a significant distance from the embers, rotating all the while. The rotation means that the heat cannot accumulate too much on any single spot: no scorching. The slow, gradual pace keeps the food on the spit tender; but the cook must also be vigilant for signs that the fire isn't hot enough or that the spit needs to be moved nearer to the fire. This is why true roasters are said to be born, not made. In addition to the sheer hard labor of rotating, you need a kind of sixth sense for the food on the spit, some instinct that forewarns you when it is about to burn or when the fire needs prodding.

It enrages Ivan Day when people say, as they often do, that spit-roasting by an open hearth—the most prized method of cooking in Europe for hundreds of years—was dirty and primitive. "On the contrary, it was frequently a highly controlled and sophisticated procedure with an advanced technology and its own remarkable cuisine." Sometimes, spit-roasting is dismissed as Neanderthal, to which Day remarked one day, warming to his theme, "I'd rather eat beef cooked the Neanderthal way" than beef prepared "in a microwave."

I have eaten several "historic" spit-roasted meats cooked by Ivan Day using his seventeenth-century fireplace and all its accoutrements. Both the flavors and textures were out-of-this-world superb. I could never be sure, though, to what extent this reflected the technology of open-fire cookery or whether it really came down to

Day's considerable gastronomic skills. His culinary standards go far beyond those of the average home cook. He candies his own citrus peel and distills his own essences. He frets over seasoning, and every meal that emerges from his kitchen looks like a still-life painting.

What all Day's spit-roasted meats had in common was a tender succulence sometimes lacking in oven-roasted meat. A leg of mutton cooked using a vertical bottle-jack emerged on the plate in deeply savory mouthfuls. Renaissance Italian veal was yielding and fragrant with green herbs. Best of all was the Victorian sirloin of beef, after a recipe by Francatelli, chef to Queen Victoria, which I learned how to make in one of Ivan's courses. First, we larded the raw sirloin. This consisted of sewing strips of cured pork fat into the meat using giant "larding needles," the idea being that it would be basted, deliciously, from within. Then, we marinated it in olive oil, shallots, lemon, and herbs—surprisingly light, Italianate flavors. Finally, we put it on a vast spit and secured it in place before the fire with metal clamps called "holdfasts." The beef was served—in high Victorian style—decorated with hatelets: skewers filled with an opulent string of truffles and prawns. The beef itself had a caramel crust from Ivan's diligent basting; the inside melted on the fork like butter. Those of us taking the course exchanged glances around the table. So *this* was why there was such a fuss about the roast beef of England. These superb results were the product of a startling and taxing range of work and equipment, which underwent centuries of refinements.

First of all, there was the fire itself. We do not know how the first fires were made, whether by deliberately striking pyrite rock against flint or lighting a branch opportunistically from a brush fire. It is certain, however, that the early domestication of fire was an anxious business: getting the fire, keeping it going, and containing it were all liable to cause problems. Paleolithic hearths (from 200,000 to 40,000 years ago) consisted of a few stones arranged in a circle to hold the fire in. At Klasies River Cave in South Africa, there are 125,000-year-old remnants of cave-dwelling humans, who seem to have eaten antelope and shellfish, seals and penguins, roasted in purpose-built stone hearths.

Once set up, a fire needs to be fueled. In places with scarce firewood, a fire might be fed with anything from turf and peat to animal dung and bone. Some hunter-gatherer tribes carried fire with them, because once the fire was out, there was no guarantee it could ever be started again. The Greeks and Romans built inextinguishable public hearths in honor of Hestia/Vesta, goddess of the hearth. Even in a domestic setting, the basic hearth fire was not lightly put out.

When we hear of an "eternal flame," we picture a neat orange fire, like that in the Olympic torch, being passed from hand to hand. But in the average premodern hut—whether Roman or Irish, Mesopotamian, or Anglo-Saxon—the eternal hearth came at the cost of marinating yourself in a foul medley of smoke and fumes. The heat in a modern professional kitchen is bad enough; I have visited the kitchens of various London restaurants for a few minutes at a time and emerged, drenched with sweat, pitying the poor commis chef who has to complete a ten-hour shift in such conditions. And these are shiny modern kitchens with all the ventilators and smoke extractors that "health and safety" require. How bad must it have been in a small ancient kitchen with zero ventilation? Near unbearable.

In the mid-twentieth century, the classicist Louisa Rayner spent some time in a wattle-and-daub earth-floored cottage in the former Yugoslavia, the kind of accommodation the great bulk of humanity lived in before the arrival of such things as basic ventilation, electric lights, and modern plumbing. Rayner suggested that this cottage was not unlike a Greek cottage from the Homeric era. The main room had no windows or chimney, only a hole in the roof for the smoke to escape. The walls were soot-black from the fire. The inner timbers were pickled all over from smoke.

Cooking in such a confined dwelling can hardly have been the kind of pleasurable activity it is for so many of us now. Every attempt to poke the dull fire or prod the half-cooked meat only adds to the smoke. You must give up hope of keeping a steady flame under the meat and open a door. No wonder many ancient Greek cooks seem to have preferred to use a portable brazier, a clay cylinder that could be moved around into any room of the house, and far more easily controlled.

Things were slightly better in the kitchens of the rich in medieval England. At least there were stone floors instead of beaten earth, and the vast high ceilings dissipated some of the smoke. Even so, while churning out the roasted meats the lords expected, the great halls of these dwellings were often choking with fumes. If cooks needed to do any additional cooking besides roasting, multiple fires needed to be built, dotted around the kitchen floor: there might be a stewing fire, a boiling fire, and a roasting fire, all ablaze, throwing off sparks and soot. In such houses, cooks were often expected to roast enough meat for fifty at a time. The danger and unpredictability of these open hearths can be gauged from the fact that English kitchens were often built as separate buildings, joined onto the hall by a covered passageway. That way, if one kitchen burned down, another could be built without disrupting the main house.

There was no question of living without a hearth, however, for without it there was no winter warmth and no roast meat. To an English patriot, the thought of a vast haunch of venison or a baron of beef slowly rotating before a fire is splendid. In the reign of Elizabeth I, someone noted that "English cooks, in comparison with other nations, are most commended for roasted meat." Englishmen prided themselves on their red-blooded tastes. "Beef and liberty!" was the cry in the eighteenth century. "When England discards Roast Beef, we may fairly conclude that the nation is about to change its manly and national character," wrote Dr. Hunter of York in 1806. To the French, we are still *"les Rosbifs."*

But the English predilection for roast beef (which in any case was largely limited to the wealthy) was not, at root, a question of taste; it was a question of resources. English cooks chose to roast great carcasses by the heat of great fires in part because—in contrast to other nations—the English were abundantly well-endowed with firewood. From the middle ages up to the nineteenth century, London was far richer in fuel than Paris, a circumstance that made the entire food supply of the English more abundant. The French may have wished that they were *"Rosbifs,"* too. Bread, beer, and roast meat were all greedy consumers of firewood; it has been calculated that simply keeping up with London's appetite for bread and beer would have taken around 30,000 tons of firewood in the year 1300, but this was no problem because there was plenty of well-stocked—and largely renewable—woodland in the surrounding counties. Still more fuel was needed to warm private homes and roast meat. After the Black Death, the cost of firewood increased dramatically in Britain, but cheap coal took its place, to keep those roasting fires roaring.

The difference with China is stark. It is true that the Chinese have their own tradition of roast meats—the windows of every Chinatown are filled with glossy whole roast ducks and racks of roast pork ribs. But wok frying remains the basic Chinese cooking technique, a cuisine born of fuel poverty. Every meal had to be founded on frugal calculations about how to extract the maximum taste from the minimum input of energy. *"Les Rosbifs"* had no such worries. The roast beef of England reflected a densely wooded landscape, and the fact that there was plenty of grass for grazing animals. The English could afford to cook entire beasts beside the heat of a fierce fire, throwing on as many logs as it took, until the meat was done to perfection. In the short run, this was a lavish way to eat; and a delicious way, if Ivan Day's re-creations are anything to go by. In the long run, it almost certainly limited the nation's cooking skills. Necessity is the mother of invention, and more restricted amounts of firewood might have forced the English into a more creative and varied cuisine.

Having enough wood didn't mean that traditional English roasting was a haphazard business. Far from it. To roast well, you needed to know which meats needed to be roasted in a gentle flame and which needed unrestricted blasting heat, such as swans. Judging from illuminated manuscripts, the know-how of spit-roasting went back at least as far as Anglo-Saxon times. Cooks needed to know how to baste the meat, in butter or oil, and how to dredge it, in flour or breadcrumbs for a crispier outside, which meant using a muffineer, a little metal shaker that resemble the nutmeg and chocolate shakers in coffee shops today. A Swedish visitor who came to England in the eighteenth century noted that "Englishmen understand almost better than any other people the art of properly roasting a joint." But once the technology was superseded, English cooks were left with an entire group of skills that couldn't easily be transferred to other cooking methods.

The key skill every English cook needed was this: knowledge of how to control a large fire, stoking it up or letting it die down, depending on the dish. A good cook knew the temperament of fire, reading patterns in the flame. To control a fire, you control the draft: by pulling air into the fire, you make the heat more intense. When Day wants to raise the temperature, he pokes it vigorously with a poker. "It will now absolutely soar!" he cries. Sure enough, ten minutes later, it is painful to go anywhere near the hearth. You feel your cheeks fry within seconds.

Cooking supper over a gentle gas flame, you can get close enough to stir and prod. Sometimes I stand with my nose over a pan, inhaling the perfume of garlic and thyme in a sauce for the sheer pleasure of it. With a roasting fire, the cooks must have kept more of a distance from the food, approaching the meat only when strictly necessary: to baste or dredge the meat or to change its position in relation to the fire. The utensils of open-fire cookery tended to be extremely long-handled: elongated basting spoons and flesh-forks, skimmers and ladles, all of which gave cooks a few extra inches of distance from the blaze. One of these long-handled instruments was the salamander, a utensil named after a mythical dragon that was supposed to be able to withstand great heat. It consisted of an elongated handle with a cast-iron paddle-shaped head. The head of the salamander was held in the fire until the iron glowed red hot, then maneuverd over a dish of food—mostly pastries, sugary creams, or dishes topped with cheese—to broil it. In the nineteenth century, this is the technique that gave crème brûlée its burned top (no need for a blowtorch). Ivan Day uses his salamander to give a crispy topping to a dish of tomatoes stuffed with breadcrumbs. He holds it a few inches above the tomatoes, and almost at once they start to bubble and brown. You can't do that with a gas cooktop.

Another critical aspect of open-fire management for a roast was getting the position of the food just right. Many people think that spit-roasting meant roasting *over* the fire, but the meat cooked a good distance to the side of the fire, only getting moved close up right at the end to brown it. This is a technique similar to a modern Argentine *asado,* a barbecue method that slow-roasts a whole animal at an angle several feet from an outdoor charcoal fire pit, until the meat is succulent and smoky. A skilled roaster knew that getting the distance right was critical for moderating the heat accumulation on the surface of the meat. Modern science has confirmed it. Recent experiments have shown that the heat intensity from a roasting fire varies by the inverse square of the distance of the meat being roasted. Each inch that you move a piece of beef nearer to the fire doesn't just make it a bit hotter; it makes it a lot hotter. With a big roast, the "sweet spot," or optimum position for roasting without charring, will be as much as three feet from the fire.

Apart from the complexity of fire, an additional problem with spit-roasting was keeping the food firmly gripped on the spit. When you stick a spit through something and turn it around, the spit has a tendency to spin while the meat stays still. Various strategies addressed this. One was to put skewer holes in the spit: the joint could be speared in place with flat skewers. Another solution was the aforementioned "holdfast," a kind of hook to grip the meat. Once the food was firmly in place, there was one more challenge facing the roaster, and it was the trickiest by far: how to keep a hulking piece of meat in perpetual motion for the hours it needed to cook.

Of all the thankless, soul-destroying jobs in a rich medieval British kitchen—scullion, washpot, drudge—there can have been few worse than that of the turnspit or turnbroach, the person (usually a boy) charged with rotating the roasting spits. "In olden times," wrote the great biographer John Aubrey, "the poor boys did turn the spits, and licked the dripping pans."

By the reign of Henry VIII, the king's household had whole battalions of turnspits, charring their faces and tiring their arms to satisfy the royal appetite for roast capons and ducks, venison and beef, crammed in cubbyholes to the side of the fireplace. The boys must have been near-roasted themselves as they labored to roast the meats. Until the year 1530, the kitchen staff at Hampton Court worked either naked or in scanty, grimy garments. Henry VIII addressed the situation, not by relieving the turnspits of their duties, but by providing the master cooks with a clothing allowance, with which to keep the junior staff decently clothed, and therefore even hotter. Turnspits were employed in lesser households, too. In 1666, the lawyers of the Middle Temple in London were making use of one "turnbroach" alongside two scullions, a head cook, and an under cook. To be a turnspit was deemed suitable work for a child well into the eighteenth century. John Macdonald (b. 1741), a Scottish highlander, was a famous footman who wrote memoirs of his experiences in service. An orphan, Macdonald had been sacked from a previous job rocking a baby's cradle and next found work in a gentleman's house turning the spit. He was aged just five.

But by that time, turnspit boys like Macdonald were something of a throwback. Over the course of the sixteenth and seventeenth centuries in Britain, their work had largely been taken over by animals. In a 1576 book on English dogs, a "turnspit" was defined as "a certain dog in kitchen service." The dogs were bred specially to have short legs and long bodies. Stuck in a wheel around 2.5 feet in diameter, suspended high up against a wall near the fireplace, they were forced to trundle around and around. The treadmill was connected to the spit via a pulley.

Some cooks preferred to use geese instead of dogs. In the 1690s, it was written that geese were better at turning spits than dogs because they kept going for longer at the treadmill, sometimes as long as twelve hours. There were signs that dogs were too intelligent for the job. Thomas Somerville, who witnessed the use of dog wheels during a childhood in eighteenth-century Scotland, recalled that the dogs "used to hide themselves or run away when they observed indications that there was to be a roast for dinner."

The turnspit breed is no longer with us. It would be nice to think that they died out because of a sudden fit of conscience on the part of their owners. But history doesn't usually work like this. Dog wheels were still being used in American restaurant kitchens well into the nineteenth century. Henry Bergh, an early animal rights lobbyist, campaigned against using dog wheels to roast meats (along with other abuses of animals, such as bear baiting). The fuss Bergh made about turnspit dogs did finally attach some shame to the practice, but it also had unintended consequences. When Bergh paid surprise visits to kitchens to check for the presence of dog wheels, he several times found that the dogs had been replaced at the fire by young black children.

In the end, it was not kindness that ended the era of the turnspit dog but mechanization. From the sixteenth century onward, inventors devised numerous mechanical jacks to rotate the spit without the need for anyone—boy, dog, or goose—to do the work. By 1748, Pehr Kalm, a Swedish visitor to England, was praising the windup iron "meat jack" as "a very useful invention, which lightens the labour amongst a people who eat so much meat." Based on his travels, Kalm claimed that "simply made" weight-driven jacks were to be found "in every house in England." This was an exaggeration. However, judging from probate inventories—lists of possessions at the time of death—around one-half of all households, not only affluent ones, did possess a windup jack, a strikingly high percentage.

Still, no wonder. Archaic as they might seem to us, these were highly desirable pieces of kitchen equipment . Mechanical jacks really were brilliant devices, culinary robots that took much of the labor out of spit-roasting. The basic mechanism was this. There was a weight, suspended from a cord, wound around a cylinder. The force of gravity made the weight slowly descend (another name for these little machines was "gravity jacks"). As it did so, the power was transmitted through a series of cogwheels and pulleys to one or more spits. Through the force of the weight dropping, the spit rotated. Some jacks rang a bell when the spit stopped.

Weight-driven jacks were not the only form of automated spit. From the seventeenth century onward, there were also smoke jacks, which used the updraft of heat from the fire to power a vane, like a weather vane. Fans of the smoke jack liked the fact that it needed no winding up and was cheap. Smoke jacks were only cheap, however, if fuel use was not taken into account. To keep the vane turning in the smoke, grotesque amounts of wood or coal had to be kept burning in the hearth. In 1800, it was calculated that you could use one-thousandth of the fuel needed to make a smoke jack work to power the spit with a small steam engine instead.

Because spit-roasting was so central to British cooking, much intelligence was lavished on inventing improved methods of turning the spit. Water, steam, and clockwork were all experimented with as ways of keeping a roasting joint in a state of constant—if not quite perpetual—motion. Mechanized spits were the gleaming espresso machines of their day: the single kitchen product on which the most complex engineering was lavished. In a seventeenth-century farmhouse kitchen, the spoons and cauldrons went back to the Romans. The spits and salamanders were medieval. The meat and fire were as old as time. But the weight-jack powering the spit was high-tech. Ivan Day still has a large collection of mechanized spit-jacks. When asked to name his favorite kitchen gadget of all time, he unhesitatingly names his seventeenth-century weight-driven jack, powered with the weight of a small cannon ball. He marvels at its efficiency. "Four hundred years before the microwave and its warning buzzer, my mechanism can tell me [when the food is done] by ringing a bell," he told BBC Radio 4's *Food Programme*. "I'd never use anything else. It works just as well now as it did 300 years ago."

In its way, the mechanized jack clearly is a miracle. It saved the pains of boys and dogs. It produces—at least in the hands of a talented cook—stupendously good roast meat, evenly cooked by continuous, steady rotation. It is a joy to watch. Few pieces of kitchenware, ancient or modern, can supply the quiet satisfaction of watching a weight-jack do its job: the speedy whirring of the flywheel, the interlocking cogs and gears, the reliable motion of the spit. On its own terms, it really works.

But technologies never exist just on their own terms. By the mid-nineteenth century, the mechanized jack was becoming obsolete, not through any fault of its own, but because the entire culture of open-hearth cookery was on the way out. Fire was in the process of being contained, and as a result, the kitchen was about to be transformed.

Ore fuel is frequently consumed in a kitchen range to boil a tea-kettle than, with proper management, would be sufficient to cook a dinner for fifty men." The author of these words was Benjamin Thompson, Count Rumford, one of the most skillful scientists ever to apply himself to the question of cooking. Among his many experiments, he set his mind to the problem of why apple pie filling tended to be so mouth-burningly hot.* Rumford was a great social campaigner, too, and believed he had found the solution to world hunger, by inventing a soup for the poor that could deliver the maximum nutrients for the minimum money. One of his other main causes was the wastefulness of roasting fires. In the late eighteenth century, Rumford was appalled by the way the English cooked over an open flame: "The loss of heat and waste of fuel in these kitchens is incredible." Rumford did not even rate the food produced by spit-roasting very highly. By focusing all their energies on roasting, English cooks had neglected the art of making "nourishing soups and broths."

---

* His answer was that the "heat passes with much greater difficulty, or much slower, in stewed apples than in pure water." Because heat conducted slower in stewed apples, it took longer for it to cool down than hot water, hence, the problem of burned mouths when eating apple pie.

Rumford's problem with English hearths was easily summarized: "They are not closed." From this basic error, "other evils" followed. The kitchen was an uncomfortable environment to work in, as anyone knew who had ever "met the cook coming sweltering out of it." The heat was excessive, there were drafts of cold air by the chimney, and worst of all, there were "noxious exhalations" from burning charcoal: a constant atmosphere of smokiness. Excessive smoke was not an accident, but inherent to the design of the English kitchen around the year 1800. To make room for all the pots that needed to be fitted over the fire, the range was built very long, which in turn necessitated an "enormously large" and high chimney that squandered much fuel and generated much smoke. Rumford's solution was his own custom-built closed range, which consumed vastly less fuel, as he had proved when he installed one in the House of Industry (the Workhouse) in Munich.

In a Rumford range, instead of one large fire, there would be lots of small enclosed ones, to minimize smoke and fuel wastage. Each boiler, kettle, or stewpan in use would be assigned its own "separate closed fireplace," built from bricks for added insulation and shut up with a door, with a separate canal for "carrying off the smoke into the chimney." The kitchen would be smokeless and highly efficient, and Rumford claimed the food produced was tastier. He summoned some friends for a taste test of a leg of mutton roasted in a Rumford roaster as against a spit-roasted leg. Everyone preferred the one cooked in the enclosed roaster, relishing the "exquisitely sweet" fat with currant jelly; or so they said.

It was one thing to convince his friends and acquaintances, still another to convince the general public. Rumford's idea was ahead of its time. His ingeniously designed stoves never found a wide audience (though various sellers would later market and sell "Rumford stoves" that had no connection with the original). Rumford's invention was not helped by the fact that it was largely made from bricks, containing very little iron. This meant that ironmongers—who at this time were the main manufacturers of cooking apparatus—had little incentive to reproduce the design.

There was also the fact that smoky and wasteful as they might be, cooks clung to their open fires as simply the only way to roast meat. Campaigners for smokeless stoves in the developing world face the same obstacles today. The average Third World open cooking fire—fueled by coal, dung, or wood—generates as much carbon dioxide as a car. Around 3 billion people—half the world's population—cook like this, with dreadful consequences, both for carbon emissions and individual health: such fires can cause bronchitis, heart disease, and cancers. The World Health Organization has calculated that indoor smoke, chiefly from cooking fires, kills 1.5 million people every year. Yet when aid workers go into villages in Africa or South America offering clean, nonpolluting cookstoves, they often encounter resistance, as people stubbornly stick to the smoky fires they have cooked on all their lives.

In 1838, four decades after Rumford's warnings on the dangers of open hearths, cookery writer Mary Randolph insisted that "no meat can be well roasted except on a spit turned by a jack, and before a clear, steady fire—other methods are no better than baking." There continued to be innovations in jack design, long after you might have expected to see the last of them. In 1845, a Mr. Norton took out a patent for a spit propelled electrically with the aid of two magnets, a strange clash of old and new technology. Over Victoria's century, Britain entered the age of gas lighting, high-speed rail travel, flushing toilets, and telephones; and still many people chose to have their meat cooked before a roaring fire. As late as 1907, the Skinners Company in London had an eleven-foot-wide roasting range installed in their Guild Hall kitchen.

The prejudice against closed-off cooking ranges was largely that they seemed too much like bread ovens. Only open fires could roast, it was believed. Ovens were things that baked. In European kitchens, the two kinds of heat were stubbornly kept apart.

In the East, this division has not existed to anything like the same extent. The Arabic word for bread is *khubz,* which generates the verb *khabaza,* meaning to bake or "to make *khubz.*" But *khabaza* can also mean to grill or to roast. This single verb thus brings together what in English would be three separate cooking techniques. All three techniques can be performed in a *tannur,* or clay oven.

Basic clay bread ovens go back at least as far as 3000 BC in the Indus Valley and Mesopotamia, on the site of modern-day Iraq and Pakistan. These bread ovens had the traditional round cylindrical clay form that they still have to this day in much of rural Africa. A fire is lit in the bottom of the cylinder, and dough is lowered in through a hole in the top and slapped on the side of the oven; it is lifted out again a couple of minutes later as flatbread. These clay ovens look like up-side-down flower pots. In Iraq, the name for these ovens was *tinaru.* We would call them *tannurs* or *tandoors,* a technology still in use throughout the Middle East, Central Asia, and Southeast Asia.

Although it has been refined over the past 5,000 years, the *tannur/tandoor* serves the same purpose it always has: a provider of intense, dry baking heat. The *tannur* enabled households, even humble ones, to be self-sufficient in bread. A series of laborers' houses have been excavated in an ancient Egyptian village, Amarna, dating to 1350 BC. Half the houses, including small ones, show traces of cylindrical clay ovens. Whereas in Europe there was a persistent belief that the only true bread was that baked by professional bakers, in medieval Iraq, homemade *tannur* bread was preferred. A market inspector in medieval Baghdad noted that "most people avoid eating bread baked in the market."

The *tannur* offered different cooking possibilities in the home than fire alone. Despite being cheap and portable, these clay ovens provided some heat control. An "eye" at the bottom could be opened or shut to increase or reduce the temperature. For some breads—such as a round Iraqi "water bread" coated in sesame oil—a more moderate heat was used. But clay ovens could also get furnace-hot when needed. Because the wood or charcoal is burned directly in the bottom of the *tandoor* and continues to burn as the food cooks, the temperatures in a modern *tandoor* can be tremendous: as much as 896°F (compared with a maximum temperature of 428°F for most domestic electric ovens). It is this blistering heat that makes this oven such a powerful and versatile piece of equipment.

The uses of the *tannur* went far beyond baking, which partly explains why in Middle Eastern and Eastern cookery, the baking-roasting dichotomy did not exist. As well as baking bread, cookies, and crackers, a *tannur* could be used for stews and casseroles and for roasting meat. Today, the *tandoor* is probably most famous as a device for cooking chicken marinated in yogurt and red spices: tandoori chicken. In tenth-century Baghdad, the *tannur* was used to roast such things as "fatty whole lamb or kid—mostly stuffed . . . big chunks of meat, plump poultry and fish." They were placed on flat brick tiles arranged on the fire or securely threaded into skewers and lowered into the *tannur* until they roasted to succulence. There was clearly no sense here that you could not "roast" meat in an oven. However, the heat of a *tandoor* works on food in a different way than a Western bread oven.

There are three different forms of cooking heat. All cooking obeys the second law of thermodynamics: heat flows from hotter things to colder things. But this transfer of energy can happen in more than one way. The first way is radiant heat. Imagine the way an Italian frittata omelette suddenly puffs and browns when you put it under the grill. The grill itself hasn't touched the omelette; and yet it is cooking. This is from heat radiation, like the sun's rays. Like radio waves, radiation works without any contact: the thing being heated and the heater do not need to touch. A red-hot fire provides lots of radiant heat, from both the flames and the embers. The moment in Ivan Day's kitchen when he poked the fire and the heat levels jumped up from bearable to unbearable represented a sudden leap in the quantity of radiant heat, enough to produce a sizzling crust on a joint of beef.

The second type of heat transfer is conduction. Unlike radiation, it works from material to material, via touch. Some materials are very good conductors, notably, metals. Others are poor conductors, such as clay, brick, and wood. When something heats up, its atoms vibrate rapidly. Conduction works by passing on these vibrations from one material to another: from a metal sauté pan to a piece of steak; from a metal saucepan handle to a tender human hand.

The third type of cooking heat is convection. It happens when the molecules in a fluid—whether air or water, stock or oil—diffuse heat to one another. The hot parts of the liquid or gas are less dense than the cold parts: think steam as opposed to water. Gradually, the hot fluid transfers energy to the cool fluid, until all is hot: think of porridge bubbling in a pot or the air in a preheating oven.

Any given method of cooking will involve a combination of these forms of heat, but one or another usually dominates. What makes the *tandoor* so unusual is that it combines all three forms of heat transfer in one. There is a massive blast of radiant heat from the fire below, plus more radiation from the heat retained in the clay walls. Bread cooked on the walls or meat cooked on skewers gets hot by conduction from the clay or the metal skewers. Finally, there is some convected heat from the hot air circulating in the oven. The *tandoor* provides intense and potent heat: the kind you can use to cook almost anything.

The ovens of Western cooking were generally brick boxes. Heat transfer in this sort of oven is typically around 80 percent by convection and only 20 percent by radiation. In place of the intense constant heat of the *tandoor* was a heat that started fierce but became progressively cooler. Indeed, the food didn't go in until the flames had already gone out. Over centuries, cooking styles evolved to reflect this gradual cooling off, with a repertoire to make the most of every phase of oven heat. Food was cooked in succession: bread went into the hottest oven, followed by stews, pastries, and puddings; later when the oven was barely warm, herbs might be left to dry in it overnight.

It is true that the West had its own equivalent of the *tannur* in the "beehive ovens" introduced by the Romans, but these never penetrated the entire food culture as the Eastern clay ovens did. In ancient and medieval Europe, bread ovens tended to be vast communal chambers, feeding an entire community with bread. The baking equipment used in a manorial or monastery kitchen was all on a giant scale: dough was stirred with wooden spoons as big as oars and kneaded on vast trestle tables. Communal baking ovens were stoked up via stoking sheds from outside. First the fuel—bundles of wood or charcoal—was heaved into the back of the oven and fired up. When the oven was hot, the ashes were raked out into the stoking sheds and the dough was shoved in, on great long wooden paddles called peels. Like turnspit boys, bakers worked almost naked because of the heat.

There the similarity ended. Western baking and roasting were entirely separate activities with separate equipment, methods, and recipes. By the eighteenth century, baking involved a paraphernalia of wooden kneading troughs, pastry jaggers, various hoops and traps for tarts and pies, peels, patty pans, wafer irons, and earthenware dishes. The baker had no need of jacks and spits, gridirons, and fire dogs. There is an engraving of the royal kitchen at St. James's Palace during the reign of George III, around the time of American independence. It depicts three different types of fire cookery. There is an open grate for roasting, a closed oven for baking, and a raised brick hearth for making stews and sauces. Each operation is entirely distinct.

No wonder Rumford's closed range met with such ridicule and derision when it was first introduced. It threatened to bring together two technologies—baking and roasting—that almost everyone in Britain, if not the Western world, deemed to be incompatible. It was as if he had said you could use a deep-fat fryer for steaming or a toaster to boil eggs.

There were also doubts from many quarters over whether the enclosed heat of an oven could ever replace the homely pleasures of warming yourself by an open fire. Could a stove whose flames were hidden from view ever be a *focus* in the way that a hearth was? A fire speaks to us in ways that are not always rational. For all the hazards and smoke of a roasting fire, those flames signified home. It was said that when stoves were first introduced in the United States in the 1830s, they inspired feelings of hatred: stoves might be an acceptable way to heat a public place such as a barroom or courthouse, but not a home.

In time, most people got over their repugnance. The "model cook-stove" became one of the great consumer status symbols of the Industrial Age, and homes developed a new focus. The typical Victorian cookstove was a cast-iron "monster" that combined a hot-water tank

for boiling and hot plates to set pots and pans on with a coal-fired oven behind iron doors, the whole thing connected with "complicated arrangements of flues, their temperature controlled by a register and dampers." By the mid-nineteenth century in both Britain and the United States, the closed range or "kitchener" had become the single essential kitchen fitting in middle-class kitchens. Cooks learned that instead of building a kitchen around a fire, it could be built around an appliance, just as today's affluent kitchens are structured around brightly colored KitchenAids and gleaming Viking ranges.

At the Great Exhibition in 1851, when Britain showed off its industrial riches to the world, many kitcheners were on display. First prize went to the Improved Leamington Kitchener, an elaborate construction, which Mrs. Beeton admired. The Leamington explicitly offered to combine the twin functions of roasting and baking with a single fire. Inside was a wrought-iron roaster with a dripping pan, but this could be converted to the unventilated heat of an oven by closing the valves at the back. The Leamington could also supply gallons of boiling water. A range was never just designed to cook food; it was needed to provide hot water for the whole household, to heat up irons and warm hands.

"Leamington" was one of the first pieces of equipment to become a household name in Britain and was soon being used as shorthand for closed ranges in general. But there were plenty of competing models, many of them with patents, glamorous names ("The Coastal Grand Pacific," "The Plantress"), and fancy squiggles and curlicues on the front. These were cooking appliances as fashion statements.

The sudden popularity of the closed range went beyond style. It was driven by the materials of the Industrial Revolution, chiefly coal and iron. There was a boom in cookstoves, not because people had read Rumford and turned against open-hearth cookery but because the market was suddenly flooded with cheap cast iron. The patent kitchen range was an ironmonger's dream: the chance to offload a great lump of iron, with added iron accessories. The rapidity with which new versions came out was an added bonus: after a couple of years, a stove might become outmoded and get traded in for a more up-to-date model, meaning more profits.

Cast-iron production had improved in the mid-eighteenth century with the discovery of a new method of production, using coal instead of charcoal. John "Iron-Mad" Wilkinson (1728–1808) pioneered the new method and produced the steam engine cylinders that hastened production even further. A generation later, cast iron was everywhere: Victorians shut themselves behind cast-iron gates, rode over cast-iron bridges, sat around cast-iron fireplaces, erected cast-iron buildings, and cooked in cast-iron kitcheners. The housekeepers and their mistresses who pored over the Smith and Wellstood catalog, wondering which model of stove to buy, may have believed that they were satisfying nothing but their own whim. But whichever fancy new design they chose, they were serving the profits of the iron industry and supporting the coal industry as well, as these new modern kitcheners were almost all fired with coal rather than wood or turf or peat.

Coal was by no means new to the British kitchen. The first coal revolution had taken place in the sixteenth century, when a shortage of wood transformed kitchens. The Elizabethan Age saw a great expansion of industry. Iron, glass, and lead manufacturing were all greedy consumers of timber. Timber was also needed for shipbuilding in the war against the Spanish, leaving far less for English hearths at home. The result was that many kitchens, particularly in towns, reluctantly converted to "seacoal," so named because it was transported by sea.

The move from wood to coal brought with it other changes. The medieval wood fire was really an indoor bonfire, with nothing but some andirons (or brand-irons) to stop the burning logs from rolling forward onto the floor. It was a hazardous form of cooking. In the seventh century, the Saxon archbishop Theodore pronounced that "if a woman place her infant by the hearth, and the man put water in the cauldron, and it boil over and the child be scalded to death, the woman must do penance for her negligence but the man is acquitted of blame." Aside from the injustice of this, it speaks of a world in which children of two or three were at high risk of toddling into hot fires and cauldrons. Women were at risk, too, because of their long, trailing dresses. Medieval coroners' reports listing accidental deaths indicate that women were more likely to die accidentally in the home than anywhere else. Little girls died at open hearths playing with pots and pans, copying their mothers.

The combination of wood-timbered houses and open hearths made kitchen fires a common occurrence. The most famous kitchen fire in British history was the blaze starting in the small hours of September 2, 1666, at the bakery of Thomas Farriner, Pudding Lane, which set off the Great Fire of London. When the city was rebuilt in brick, the new houses had coal-burning grates.

One of the effects of a switch to coal was to enclose the fire—at least a little. Coal needs a container, in the form of a metal grate, called a "chamber grate" or "cole baskett." The switch from down-hearth wood fires to grated coal fires was accompanied by a whole new battery of equipment. The new fires needed cast-iron firebacks to protect the wall from the fierce heat and complex fire cranes to swing pots over the fire and off again. The other great change brought about by coal was the chimney. The great increase in chimneys in Elizabeth's reign resulted largely from the increased use of coal, because wider flues were needed to carry away the noxious fumes of the coal as it burned. In fact, as Rumford observed, this combination of very wide chimneys and blazing roasting fires was deadly. When Pehr Kalm arrived in London from Sweden in the eighteenth century, he found the "coal-smoke" from cooking "very annoying," and wondered if it was responsible for the high incidence of lung disease in England. He developed a terrible cough, which only abated when he left the city.

Not everyone switched to coal. In the countryside and in the northern counties, the norm remained the old down-hearth wood fire. Meanwhile, the poorest families in both city and country muddled by as best they could with whatever fuel was at hand: handfuls of dry heather, twigs gathered from the hedgerows, cattle dung. Not for them the shiny new patent cookstoves.

It is debatable whether being unable to afford a coal kitchener was a great loss. The closed range in this particular form had many disadvantages and few real benefits over an open fire. Unlike Rumford's ideal closed hearths built of brick, many early ranges were badly constructed, belching coke fumes. A letter of 1853 to the *Expositor* called them "poison machines," drawing attention to the recent deaths of three people from inhaling their fumes. And besides that danger, many of the ranges were inefficient. Promoters of American cookstoves claimed they would save around 50–90 percent of fuel compared with an open hearth, but this did not take into account the heat wasted. A good stove needs to insulate heat as well as conduct it. There was a fundamental problem in using all that highly conductive iron, which absorbed vast amounts of heat and then radiated it back out into the kitchen rather than into the food, leaving the poor cook in a furnacelike atmosphere of heat, ash dust, and soot.

The cast-iron kitchen range was one of those curious technologies that became an object of consumer desire without offering much real improvement on what came before. It didn't save labor—quite the opposite, in many cases. Getting a fire started was no easier in a stove than on a hearth, and polishing and cleaning the range was practically a full-time job, whether for a servant or a wife. As late as 1912, a housewife married to a policeman listed her daily duties relating to the range:

1. Remove fender and fire-irons.
2. Rake out all the ashes and cinders; first throw in some damp tea-leaves to keep down the dust.
3. Sift the cinders.
4. Clean the flues.
5. Remove all grease from the stove with newspaper.
6. Polish the steels with bathbrick and paraffin.
7. Blacklead the iron parts and polish.
8. Wash the hearthstone and polish it.

All this work and not a single dish has been prepared; not a rasher of bacon has been fried, not a potato boiled. Unlucky woman. If only she had been born a few years later, she might have been spared it all. She would almost certainly have gotten a gas oven.

Our domestic lives are all composed of hundreds of small, daily, recurring activities, nowhere more so than in the kitchen. The devices that are truly revolutionary are not the ones that enable us to make entirely new creations—air-drying strawberries or vacuum-cooking rare cuts of venison—but the ones that let us do the things we already do with greater ease, better results, and more pleasure: making family breakfast more speedily for less money and far less trouble, for example. The gas range was a rare breakthrough: a tool offering real progress in the kitchen.

Compared to a coal range, gas was cleaner, pleasanter, and cheaper: it was estimated that for a middle-class English family, the cost of cooking with gas was around 2-½ pence a day, as against 7 pence to 1 shilling for coal. The real joy of gas, however, was the work it saved. The early cooks who learned to prepare meals using gas in the 1880s went into rhapsodies over how much easier life had become. A simple job such as cooking the morning breakfast took far less "time and attention" than under the old system. Mrs. H. M. Young, who wrote one of the first cookbooks to include a section on gas, noted that "a breakfast for a medium family, say, coffee, chops, steaks or bacon, eggs and toast, may be prepared easily in 15 minutes."

As is so often the case, the innovation was initially met with suspicion and resistance. There was a time lag of nearly a century between the first experiments in gas cookery and its adoption by a wider public. The same cooks who toiled in the tropical heat and filth of a coal-fire range feared that gas was a dangerous form of cooking, which would make the food taste and smell disgusting. Although increasingly happy to light their homes with coal gas— London was the first city to be lit by gas, in 1814—people feared they would either be poisoned or die in an explosion if they cooked with it. Servants were said to be scared senseless by gas ovens.

Perhaps some of the prejudice was justified, insofar as the earliest models of gas oven were badly ventilated and the burners did not always give an even flow of gas, which did indeed result in some gassy-tasting food. But the prejudices continued long after gas cookery had become safe and reliable. Ellen Youl, a working-class housewife from Northampton, acquired a gas stove at the end of the nineteenth century. Ellen's husband reacted with horror.

He thought the gas contained poison and refused to eat anything cooked by it. Ellen, however, would not get rid of her new labour-saving contrivance. She cooked his dinner every day in the gas stove, transferring it to the open fire a few minutes before he returned from work.

The very first experiments in gas cooking had an element of scientific showmanship, as if to highlight the novelty. The first commercial gas cooking apparatus sold in Britain appeared in 1824, produced by the Aetna Ironworks. It looked a bit like a horizontal squash racket, fashioned from gunmetal and pierced with holes, through which the gas jets flowed, creating open flames. There was no surrounding oven or cooktop: you just placed it under whatever you wanted to cook to create a heat source, like a Bunsen burner. It would be another half a century before gas cooking became widespread, despite the attempts of Alexis Soyer, the Victorian celebrity chef, who marketed a very expensive and fancy gas stove called the Phidomageireion that boasted the impossibility of "explosion ever taking place." This didn't entirely reassure. Many people must have shared the opinion of Thomas Webster (author of the *Domestic Encyclopedia*) in 1844 that gas cookery was simply "an elegant culinary toy," an addition to the "usual means of cooking" rather than a replacement for it.

It was only in the 1880s that manufacturers—notably, William Sugg, whose family cornered the market in gas stoves for some time—finally started producing equipment accessible enough to convert the staunchest coal-range user. Sugg gas ranges looked remarkably like coal ranges, and they came with the same type of fanciful names: the Westminster, the Cordon Bleu, the Parisienne. Reassuringly, for lovers of an old English roast, meat cooked in the oven was still suspended over a dripping pan, reminiscent of an old open fire. The Sugg company came up with a good solution to quell the fear of explosions, fitting all the burners with flash lights to light them with the turn of a knob, avoiding the need for matches.

The 1880s also saw the spread of the penny-in-the-slot gas meter, which made gas cookery affordable for all but the very poorest in areas with gas supplies. Gas companies installed the meters free of charge and also rented out ovens for a modest quarterly cost. Subscriptions increased rapidly. In 1884, the Newcastle-upon-Tyne and Gateshead Gas Company rented out just ninety-five gas stoves; by 1920, the number had increased to 16,110. By 1901, one in three British households had a gas stove; by 1939, on the brink of World War II, three-fourths of households cooked with gas. In other words, the majority of people were finally liberated from what had been one of the defining activities of human life, the business of starting and maintaining a fire.

By this point, gas ovens had a challenge in the form of electricity. Thomas Edison first created a successful light bulb in 1879, but electric cooking was much slower to take off, hampered by the expense of early electric stoves and the limited availability of an electricity supply. The Science Museum in London has in its collection the earliest surviving electric oven, consisting of a cookie tin joined to a large light bulb with some coils of wire. It does not look very promising. In 1890, the General Electric Company started selling an electric cooking device, claiming it could boil a pint of water in twelve minutes—which only serves to bring home just how slow much cooking was in the era of coal fires.

Electric cooking only became normal—both in Europe and the States—in the late 1920s, as the price of electric cookers decreased and their efficiency improved. Early electric ovens took ages to preheat—as much as thirty-five minutes in 1914—and the heating elements had a tendency to burn out. And they were expensive both to buy and to run. An average family might buy an electric kettle or toaster but had little incentive to upgrade a gas oven to an electric one. The electric refrigerator performed functions that simply did not exist before. The electric oven was less revolutionary (its only real advantage before the invention of inbuilt safety devices to cut off unlit gas was that you couldn't asphyxiate yourself in one). Its great benefit—of providing cooking heat that could be switched on and off at will—had already been achieved by the wonders of gas. By 1948, 86 percent of households in Britain used electricity in some form. But only 19 percent owned an electric stove.

Now, like many, I cook using a combination of gas and electric. My oven is an electric convection type (using a small fan to circulate air better) with a separate grilling oven on top. It does the job OK. I put in flat cake mixture; it comes out risen. It roasts potatoes evenly enough, and I can peer through the glass door to make sure nothing is scorching. But I feel nothing like the same affection for it that I feel when cooking at my gas cooktop, which offers all the benefits of fire and none of the drawbacks. The few times I have cooked on an electric induction cooktop, it has driven me to despair: the flat surface, an invitation to chubby little fingers to burn themselves. One minute it is stone cold, then suddenly and seemingly without warning it is red hot (though admittedly I haven't used the very latest generation of induction cooktops, which are currently being trumpeted as the last word in efficient heat). Gas does my bidding. When I hear that *click-click-click*, waiting for the flame to light, I know good things will happen. In 2008, the Chinese food writer Ching-He Huang offered some sound advice on wok cookery to people who did not have a gas cooktop: "Invest in a new stove!"

Apart from the original invention of cooking, gas-powered heat was the single greatest improvement ever to occur in kitchen technology. It liberated millions from the pollution, discomfort, and sheer time-wasting of looking after a fire. A further step away from the open hearth came with the microwave oven, though this time the benefits—both culinary and social—were less straightforward. Today, with new markets in China just opening up, global microwave sales stand at around 50 million a year. In many small city kitchens the world over, a microwave is the major way of applying heat to food. Cooks clearly do a lot of microwave cooking. Yet it remains a controversial tool that has never inspired the love we once felt for fire.

The microwave is not always given enough credit for the many things it does exceptionally well. It can cook fish so that it stays moist and make old-fashioned steamed puddings in minutes. It is a nifty device for caramelizing sugar with minimal mess and for gently melting dark chocolate without it seizing up. It cooks perfect fluffy Basmati rice effortlessly. The attraction of microwaves for fat molecules makes it the ideal way to de-fat ducks and spare ribs before roasting, as Barbara Kafka notes in her 1987 opus, *Microwave Gourmet*, the most persuasive case ever made for the microwave as an instrument of pleasure.

Yet the microwave is just as likely to inspire thoughts of panic as pleasure. These "fireless ovens," as they were initially called, seemed baffling objects when they were first sold in the 1950s, and many cooks remain baffled and alarmed by them today. The invention came in 1945 from Percy Spencer of the Raytheon Company, an engineer who was working on military radar systems, trying to improve the magnetron, a vacuum tube for generating microwaves. Various mythical stories are told of the moment when Spencer first noticed that the magnetron generated enough heat to cook. In one version, he was leaning against an open waveguard—the tube through which waves travel—when he noticed that the chocolate bar in his pocket had melted. Others say he stared, amazed, as an egg exploded and cooked itself; or that he left his lunchtime sandwich on the magnetron and returned to find it cooked. The team of engineers who worked alongside Spencer later said that the truth was less dramatic: it took a series of methodical observations by several people rather than one man's eureka moment to hatch the microwave oven. However it happened, it took a huge leap of imagination on the part of Spencer and his team to think that the magnetron, this vast metal cylinder, could be used not in the field of war but in a kitchen. The QK707 magnetron used in very early models weighed a colossal 26.5 pounds, as against the 1.5 pounds of a standard modern microwave. Spencer showed still more imagination in realizing immediately what would be one of the microwave's most popular uses: making popcorn. An illustration on Spencer's second microwave patent showed how an entire ear of husked corn could be seasoned with butter and salt, placed in a waxed paper bag and turned into popcorn in just "20 to 45 seconds." In 1945, this seemed highly unlikely; and indeed it would be another two decades before the domestic microwave oven was a mainstream proposition (sales only took off in 1967, when manufacturers managed to get the price of an oven below $500).

Many consumers still find the microwave an unlikely way to cook. It seems a step too far away from fire to be anything good. For a long time it was feared on health grounds. It is true that older models sometimes leaked more than 10 mW/cm² of radiation, compared to the new, extremely stringent standards of 1 mW/cm². But in either case, it was vastly less "radiation" than you would be exposed to simply by standing around two feet from a fireplace (50 mW/cm²). Based on all the evidence to date, the microwave is innocent of health hazards, beyond the dangers of cooking with it, such as small objects exploding in "hot spots." You can avoid most of these just by reading the instruction manual.

Lying behind the periodic microwave health scares is a more fundamental suspicion of the device as a method of cooking. In 1998, a Mintel market report on microwaves in the UK found that 10 percent of consumers doggedly insisted that they would "never buy a microwave oven." Until very recently, I was part of that 10 percent. I was age thirty-six before I acquired my first one, having been brought up to believe that there was something weird about cooking "from the inside out." In my family, we viewed the microwave as only slightly less malign than nuclear bombs. How could "zapping" food possibly result in anything good?

Microwave cooking seems inexplicable in a way that other cooking methods do not. This is unfair. Microwaving does not in fact cook from the inside out as I'd always been told. There is nothing paranormal about it. Microwaved food obeys the same laws of physics as a spit roast. Microwaves travel quickly, but they only penetrate food by around 4 to 5 cm (which is why small pieces cook best in a microwave). Fat, sugar, and water molecules in the food attract the microwaves, causing them to jump around very fast. These vibrations produce heat within the food. Beyond 4 to 5 cm, the heat spreads by conduction to the rest of the food—just as it would in a frying pan. Unlike in a frying pan, where food develops a lovely golden crust, microwaved food does not brown (though some models have browning functions to compensate).

You cannot roast in a microwave, nor make bread. But no cooking tool can do everything, no matter what the manufacturers may say. It is no more of an argument against the microwave to say it can't roast than to say of a bread oven that it's too hot for making custard. The real drawback of the microwave is not with the device itself but with how it is used. The microwave had the misfortune to be first marketed in the era of postwar convenience food. "Reheating" rather than cooking food is the most common use of the microwave, according to a 1989 UK market report: 84 percent of households used it for reheating precooked food, whereas 34 percent used it for all cooking. "I don't actually cook with it," said one focus group participant, "just warm things up." In most kitchens, the microwave is not used as a form of cooking, but as a way of avoiding cooking, by slinging a frozen precooked meal in and waiting mindlessly for the beep. The microwave provided a way to eat hot food without the sociability of sitting around a family table. Most microwaves are not big enough to cook for more than one or two at once.

Is it the end of social life as we know it? Historian Felipe Fernández-Armesto excoriates the microwave as a device "with the power to change society" in a malign way, by returning us to "a pre-social phase of evolution." It is as if we never discovered fire. Throughout history, we have sought to enclose and control fire, this focus of our social lives. We tamed it with rock hearths; we built great halls around it; we enclosed it in metal grates; we shut it off in cast-iron ranges; we submitted it to our will with the gas oven. Finally, we found a way of cooking without it in the microwave.

There are signs that we miss fire and regret its absence from our lives. The enthusiasm with which amateur cooks whip out their barbecues at the first hint of sun, singeing sausages over the fire, implies that perhaps our cooking has lost its focus. No one sits around a microwave telling stories deep into the night. Its angular glass frontage cannot warm our hands or our hearts. Perhaps all is not lost, however. The process of cooking has a power to draw people together even when it does not follow the conventional old patterns. Those who believe that a microwave cannot be a focus for a home like the old hearth have never seen a group of children, huddled together in silent wonder, waiting for a bag of microwave popcorn to finish popping, like hunter-gatherers around the flame.

～ *Toaster* ～

MAKING TOAST IS SATISFYING. YOU COULD say that's because it's such a comforting substance—the crispness, the heavenly aroma as yellow butter slowly melts into the crevices. But the satisfaction is also mechanical and childish: fitting the slices in the slots, setting the timer, and waiting for a ping or a pop.

For something so basic, the electric toaster arrived late. From the 1890s, gadget-crazy late Victorians could in theory use electricity to boil kettles and fry eggs, yet for toast they still relied on the toasting forks and gridirons of open-hearth cookery. These were variations on the theme of prongs and baskets for holding bread (or morsels of cheese and meat) before the flame. Toasting, when you think about it, is really roasting: applying dry radiant heat to something until the surface browns.

Before the electric toaster could be invented, it was necessary to find a durable metal filament strong enough to withstand roasting heats without fusing. That came in 1905 when Albert Marsh discovered Nichrome, a nickel-chromium alloy with low conductivity. Then, the US market became flooded with electric toasters. There were Pinchers, Swingers, Flatbeds, Droppers, Tippers, Perchers, and Floppers: the names refer to different manual techniques for expelling the toast.

The toaster as we know it was the invention of Charles Strite, a mechanic from Minnesota fed up with the burned cafeteria toast at work. In 1921, Strite was granted a patent for a toaster with vertical pop-up springs and a variable timer. Here was something new: a toaster that could be left to do the job itself. "You do not have to Watch it—The Toast can't Burn," insisted an ad for Strite's Toastmaster. If only. It's still possible, alas, to make burned toast in a pop-up toaster.

4

MEASURE

Count what is countable, measure what is measurable,
and what is not measurable, make measurable.
GALILEO GALILEI, 1610

Don't ask me to count the hundreds and thousands.
NIGELLA LAWSON

ANNIE MERRITT FARMER WAS A COOK WHO HATED
sloppiness in the kitchen. She had no truck with a handful of this and a pinch of that, preferring to deal in fixed
and level measures. Her magnum opus, *The Boston Cooking School Cookbook* (1896), was the best-selling American recipe
book of the early twentieth century; by 1915, it had sold more than
360,000 copies. Much of its appeal came down to its insistence—so
comfortingly scientific—on using correct and accurate measurements in cooking. "A cupful," wrote Farmer, "is measured *level* . . . A
tablespoonful is measured *level*. A teaspoonful is measured level."
Farmer, a stout red-haired woman, used the same words in her
cookery demonstrations, always employing a case knife to level off

the top of the measure. No stray grains of flour would be allowed to intrude on Farmer's pastry. Her nickname was the "Mother of Level Measurements."

Farmer believed she was leading America into a new era of accuracy in the kitchen. Gone were the dark ages of guesswork. "Correct measurements are absolutely necessary to ensure the best results," she wrote. Measuring is a way of imposing order on a chaotic universe. Fannie Farmer was not just teaching her middle-class readers how to cook; she was offering them a feeling of absolute control in the domain of the kitchen. It is odd, then, that Farmer should have chosen a method of measuring—the cup system—that is so erratic, ambiguous, and prone to wildly fluctuating results.

The cup system quantifies all ingredients, whether wet or dry, fluffy or dense, by using measuring cups of a certain volume: 236.59 ml, to be precise. Because the system measures volume rather than weight, it is sometimes referred to as "volumetric." Cup measures are still nearly universally used in American cookbooks, and therefore in American kitchens, even though there are frequent complaints that measuring by weight using scales would be far easier and more accurate. Through a strange quirk of history, the United States is the only country in the world that measures food like this. Cooks in Australia and New Zealand dip in and out of the cup measuring system, and Europeans generally use volume to measure liquids, but only in the United States is this very specific unit of volume seen as the default way to measure all ingredients—animal, vegetable, or mineral—and this is due in large part to the lingering legacy of Fannie Farmer.

Fast-forward to the present day when, one summer's evening, I am attempting to cook from one of Fannie Farmer's supposedly infallible recipes. It sounds simple enough: String Bean Salad.

Marinate two cups cold string beans with French Dressing. Add one teaspoon finely cut chives. Pile in center of salad dish and arrange around base thin slices of radishes overlapping one another. Garnish top with radish cut to represent a tulip.

Have you ever tried to cram finely cut chives into a teaspoon and level the top with a knife? Don't. The chives fall everywhere. It would make far more sense just to snip them directly into the dish; a few more or less wouldn't matter. As for measuring out two cups of "cold string beans," it is a joke. The beans poke about at angles, making the task impossible. To get two perfect level cupfuls of cold green beans, you would have to crush them so much that the salad would be ruined. The recipe is also notable for the important things it doesn't tell us: how much French dressing? How long do you cook the string beans before you make them "cold"? How do you trim them? And where do you lay your hands on a radish "cut to represent a tulip," because I'm certainly not making one ("begin at root end and make six incisions through skin running three-fourths length of radish," instructs Farmer, dauntingly). There is a lot more to a recipe than measurement. Equally, however, no recipe has ever measured all the possible variations of cooking. With her faith in cups, Fannie Farmer thought she had measuring all sewn up. But the truth is that it never is.

Such is the story of measuring in the kitchen. Good cooking is a precise chemical undertaking. The difference between a truly great dinner and an indifferent one might be thirty seconds and one-quarter of a teaspoonful of salt. Recipes are an attempt to make dishes reproducible. In science, reproducibility is the ability of an experiment to be accurately re-created by an independent researcher. This is exactly the quality we seek in a recipe: your recipe for apple pie should ideally taste the same when I make it at home in my own kitchen. But cooks work in conditions with far more extraneous variables than any scientist would allow: unreliable oven temperatures, changeable raw ingredients, not to mention cooking for audiences with different tastes. The cook who gets too hung up on measuring for its own sake can lose the meal for the measuring cups. Focusing on an exact formula can make you forget that the best measure any cook has is personal judgment.

It's also worth remembering that tools of measurement in the kitchen can be judged by more than one criteria. The first is accuracy:

whether your measurement corresponds to a fixed value. Is the pitcher you are using to measure a liter of milk really a liter? The second is precision: the refinement of your measure. Could you measure out milk to within half a milliliter? The third is consistency (scientists would say reproducibility): an ability to measure out the same liter of milk over and over again. The fourth is convertibility: the extent to which a measure fits understandably into a wider system of weight and volume, and whether the tool and the units you use for measuring milk could be used to measure other things, too. The fifth may be the most important. It is ease (or user-friendliness): an ability to measure a liter of milk without any great ceremony, resources, or skill. Judged by this final criterion, one of the greatest measuring tools is the modest Pyrex measuring cup. As well as having nice clear graphics, showing both metric and imperial measures, the Pyrex cup, made from heatproof glass first patented in 1915, has a pourable spout, can withstand both freezers and microwaves, and has an invaluable ability to bounce when dropped, so long as the kitchen floor is not too hard.

All cooking entails measuring, even when it is only the spontaneous calculations of the senses. Your eyes tell you when sautéing onions are translucent enough. Your ears know when popcorn has finished popping. Your nose tells you when toast is about to burn. The cook is constantly making estimates and decisions based on those calculations. Volume and time, temperature and weight: these are the variables every cook has to navigate. But attempts to measure these things with greater accuracy through superior technology have not always led to better cooking. A fixation on formulas in the kitchen can become counterproductive. No technology has yet supplanted the measuring capabilities of a good cook, blessed with a sharp nose, keen eyes, asbestos hands, and many years at a hot stove, whose senses appraise food more tellingly than any artificial tool.

O f all the things that identify us as Americans and set us off from other peoples," wrote the great food critic Ray Sokolov in

1989, "the least ambiguous—and the one most seldom noticed—is the measuring cup." Sokolov noted that nowhere else but in the United States does an "entire nation habitually and almost exclusively measure dry ingredients with a cup." The rest of the world measures flour (at least most of the time) by weight.

Scales assume many guises, but the principle is always the same: to measure weight.* To this end, a French cook might use a balance beam with a shallow tray in it, the kind used elsewhere to weigh newborn babies. In Denmark, the kitchen scale may be an unobtrusive circle attached to a wall, looking rather like a clock, a tray cleverly flipping down to reveal a dial. The English are still fond of old-fashioned Queen scales—the classic mechanical balance scale made from heavy cast iron, with a brass dish on one side and a set of gradated weights on the other. Or maybe that's just me. Come to think of it, when friends see my kitchen, they often exclaim over the scales as if they were a museum piece, sometimes asking whether I actually use these antiquated weights. Yes! Every day! Although not, I must admit, when accurate weighing is critical, because then—obviously—I measure digitally. Cooks everywhere in the developed world now use digital scales. These are one of the best tools in the modern kitchen, offering great accuracy and precision for little money. On scales that have a zero function, you can even weigh ingredients straight into a mixing bowl, setting the scale back to zero after you put the bowl on the scales. This saves on cleanup

---

* Technically, when we say "weight" we really ought to say "mass." Weight refers to the force exerted on an object by gravity (w = mg, where m = mass and g = gravity). As such, the weight of a cup of flour on the moon would be very much less than on earth. Mass, by contrast, remains constant, regardless of environment: 100 g of flour is always 100 g. This is actually what we mean when we talk of "weight." However, because this book is concerned with practical technology rather than pure science, I will continue, inaccurately, to use the word "weight" as it is commonly understood, as a synonym for mass.

and is especially useful for things like syrup and honey, because there's no need to scrape it messily from the scales to the bowl.

But some of the older methods of weighing worked remarkably well, too (albeit with a larger margin of error). If you are German and of a traditional disposition, it is possible that you have a seesaw balance with a cup for ingredients at one end and a counterweight at the other, with weights printed on the beam. You slide the beam until it balances perfectly, then check the weight marked on the beam. This is a technology—if flimsier in construction—identical to a metal steelyard balance found at Pompeii, dated 79 AD.

The science of weighing things has been largely solved for 2,000 years or more. The oldest Chinese balance goes back to the fourth century BC: the classic design of two pans suspended from a pole. This is not to say that many people could afford their own set of scales. At the start, scales were used for weighing precious things like gold; they crept into the kitchen only centuries later. They were certainly there by the time of Apicius, author of the first ancient Roman "cookbook," who talks of the weights of not just dry ingredients ("6 scruples of lovage") but wet ones, too ("1 pound of broth"). So the technology of weighing ingredients has been established for a very long time. And it works better now than ever. Most digital kitchen scales can weigh an ingredient within an accurate range of a single gram. The wonderful thing about weighing is that you do not have to worry about density: 100 g of brown sugar is still 100 g, whether it is tightly packed or fluffy. All that matters is the weight, which is constant. It's like the old joke: which weighs more, a pound of gold or a pound of feathers? Of course, they both weigh the same. A pound is a pound is a pound.

By contrast, the American volumetric cup measurement system, at least as applied to dry ingredients, can be maddeningly imprecise. A cup of something is not just a cup. Experiments have shown that a cup of flour may vary in weight from 4 to 6 ounces, just by changing the degree to which the flour is sifted and airy or tamped down. This makes the difference between a cake that works and one that doesn't; between batter that is far too thick and batter that is watery and thin. Let's assume that the recipe writer wants you to use a "cup" of flour that corresponds to 4 ounces, but you instead measure one that comes in at 6 ounces. You will end up with one and a half times the flour required: a huge imbalance.

The problem with using volume to measure solid materials is that of compression and expansion. Under normal conditions, and assuming that it is not freezing or boiling, the density of water is fixed; you cannot squash it smaller. Flour, by contrast, may be compressed tightly into the cup or fluffed up with lots of air. Some recipes attempt to get around this by stipulating that the flour must be sifted before it is measured, and some even go into detail about exactly how much sifting should take place; but this is still no guarantee of accuracy, because flours vary so much. The sifting also adds a labor-intensive extra step to the recipe. The cook dances around with sieves and spoons, fluffing and packing and heaping and sifting, all to achieve less accuracy than a pair of scales could provide in seconds.

Moving beyond flour to other substances, cup measurement can be more maddening still. It is one thing to measure grains such as rice and couscous and porridge oats in a cup—indeed, this is probably the best way, because you want to be able to gauge the ratio of grains to cooking water by volume; the absolute quantity is less important. The ratio for porridge and most types of rice is 1:1.5, solid to liquid; couscous is 1:1. There's a certain satisfaction in pouring a measure of couscous into a measuring cup, then pouring it out and trying to get the water or stock up to the same level. You are retracing your own steps. It is something else altogether to try to measure out 5 cups of cubed zucchini (the equivalent of a pound) or 10 cups of bite-sized lettuce (again, a pound). How do you manage the chopping? Do you cut the vegetables piece by piece, adding each one to the measuring cup as you go along, or do you do it all in one go, risking chopping too much? Do you tamp the cubed vegetables down into the cup, or do you assume that the cookbook author has allowed for the spaces in between? Or do you fling the cookbook on the floor in fury at being asked to perform such an absurd task?

America's attachment to cups really is odd (and indeed there are finally small hints of rebellion against it, such as a *New York Times* article from 2011 making a "plea on behalf of the kitchen scale"). In countless ways, America feels like a more rational place than Europe. Most American city streets are laid out in orderly numbered grids, not piled up higgledy-piggledy as in London or Rome. Then there is the dollar, in use since 1792, an eminently reasonable system of currency. When it came to money, America established a usable system much sooner than Europe (with the exception of France). In the mid-twentieth century, the process of buying a cup of coffee in Rome using Italian lira was an exercise in advanced mathematics; it was not much better buying a pot of tea in London, as the British clung to the messy system of pounds, shillings, and pence. Meanwhile, Americans strolled to the grocery store and easily counted out their decimal cents, dimes, and dollars. Likewise, American phone numbers are neatly standardized in a ten-digit formula. An American friend describes the method or lack of it governing British phone numbers as "a baffling hodgepodge." So why, then, when it comes to cooking, do Americans throw reason out of the window and insist on measuring cups?

American cup measures can only be understood in the context of the history of weights and measures. Viewed historically, an absence of clear standards in measures has been the rule rather than the exception. Moreover, cup measures belonged to a wider system of measurement, within which they made considerably more sense than they do today. Our present confusion has its roots in medieval England.

A pint's a pound the world around" goes the old saying; and so it was at one time. In the Anglo-Saxon period, the "Winchester measure" was established in England, Winchester being the capital city then. This system created an equivalence between the weight of food and its volume, which would have been an obvious way to create units of volume where none had existed before.

Think about how tricky it would be to establish the exact capacity of a vessel if you didn't have a measuring vessel. How could you say how much water a given glass held? You could pour it out into another glass and compare the level between the two. But then how would you know how much the second glass held? The exercise quickly becomes nightmarish. It was much easier to establish given capacities by using the volume of certain known, weighed substances. A "Winchester bushel" was defined as the volume of 64 pounds of wheat (which was relatively constant, wheat grains being less variable in density than flour). A bushel was made up of 4 pecks. A peck was 2 gallons. A gallon was 4 quarts. And a quart was 4 pints. The upshot of all this was a pleasing fact: the Winchester bushel came in at 64 pounds (of wheat) and also 64 pints (of water). A pint really was a pound. Neat.

If only these Winchester measures had been the sole standard for volume, all would have been well. But in medieval England, numerous competing gallons came into use for different substances. As well as the Winchester gallon (also known as the corn gallon), there was the wine gallon and the ale gallon, all representing different amounts. The ale gallon was bigger than the wine gallon (around 4.62 liters as opposed to 3.79 liters), as if reflecting the fact that ale is usually drunk in bigger quantities than wine. It's all too easy to succumb to this kind of unhinged logic when thinking about how to measure things. It's like Nigel, the rock star in the film *This Is Spinal Tap*, who believes that to make music louder you need to create an amp that goes up to eleven instead of ten.

The lack of standardized weights and measures was a problem for customers wanting to receive their due (a pint of ale varied hugely from county to county) but also for the state, because it affected the duty charged on goods. The Magna Carta of 1215 attempted to address the lack of uniformity: "Let there be one measure of wine throughout our whole realm; and one measure of ale; and one measure of corn." This didn't work; competing measures continued to proliferate. Between 1066 and the end of the seventeenth century, there were more than twelve different gallons, some assigned to solids and some to liquids.

By the late eighteenth century, there were various moves to escape the anarchy of the medieval measuring system. In the 1790s, after the French Revolution, the metric system began to be established in France. The meter was based on the findings of an expedition of scientists to measure the length of the earth's meridian, an imagined line from the North Pole to the South Pole: a meter was supposed to be one ten-millionth of the meridian, though due to a tiny miscalculation it is actually a little smaller. But the principle was now set that the French would measure in tens. In 1795, the new measures were decreed in a law of 18th Germinal: liters, grams, meters. Sweeping away the old jumble of archaic standards was meant to demonstrate how modern France had become; how rational; how scientific; and how commercial. Everything, from street systems to pats of butter, was subdivided into perfect tens. The revolutionaries even experimented with a ten-day week—the "*décade*." Thanks to this new measuring system, life was now logical. You breakfasted on bread measured in grams; you drank your coffee in milliliters; you paid for it in decimal francs and sous.

The Americans and the British conducted their own reforms, but neither country wished to go as far as the revolutionary French. In 1790, President George Washington gave his secretary of state, Thomas Jefferson, the job of devising a plan to reform weights and measures. The United States already had decimal coinage, having thrown off pounds, shillings, and pence along with the British crown. But in the event, Congress couldn't agree on either of Jefferson's proposals for reform and spent several more decades failing to decide anything on the matter.

Meanwhile, in 1824, the British acted. There was no question at this time of following the French—national enemies against whom the country had only recently ceased to be at war—down a route of total metrification; the aim was simply freedom from the dark ages of multiple standards. In 1824, parliament voted to use a single imperial gallon for both dry goods and liquids. The new British imperial gallon was defined as "the volume occupied by 10 pounds of water at specified temperature and pressure." This worked out as 277.42 cubic inches, which was close to the old "ale gallon." Once the new gallon had been established, it was easy to readjust the pint, quart, and bushel measures to fit. The adage now went like this:

> A pint's a pound the world round.
> Except in Britain where
> A pint of water's a pound and a quarter.

For Britain, read the British Empire. These new imperial measures were confidently promulgated wherever the British ruled. A pint of maple syrup in Colonial Canada was the same volume as a pint of whiskey in Colonial India.

Did this spell an end to confusion in measuring? Not at all. In 1836, the US Congress finally established American uniform standards and decided to take the opposite route to Britain. Instead of adopting the new single imperial gallon, the United States stuck with the two most common gallons from the old system, the Winchester (or corn) gallon for dry goods and the Queen Anne (or wine) gallon for liquids. It is not so surprising that America wanted different standards from Britain. The strange part is that the United States expressed its metrical freedom from Britain, not with its own modern measures but with quaint old British ones. When America sent a man to the moon, that man was still thinking in the pints and bushels of eighteenth-century London. Even now, in this age of Google searches, when the cook of the household is more likely to search for a recipe online than to scour the pages of *The Joy of Cooking*, the recipes flickering on the screen of American cooking websites are still overwhelmingly given in traditional cups.

The result has been nearly two hundred years of mutual non-comprehension in the kitchen between the two nations, made still worse since 1969 when Britain finally officially joined the metric nations (though many a British home cook still prefers imperial). The United States is today one of only three countries not to have officially adopted the French metric system. The other two are Liberia and Myanmar (Burma). To American ears, there is something cold, inhuman almost, about the European practice of quantifying ingredients in grams. To the rest of the world, however, American cups are plain confusing. How much is a cup, anyway? In Australia, the cup has been defined metrically, as 250 ml. But in the United Kingdom, it is sometimes translated as 284 ml, half a British pint. Canada weighs in with a 227 ml cup, corresponding to 8 imperial fluid ounces. As for the true American cup, it is none of these. The technical definition of a US cup is half a US pint, or 236.59 ml.

Given all this, why did Fannie Farmer in 1896, the "Mother of Level Measurement," deem the cup system to be so superior and so exact? There was nothing inevitable about America's preference for volume measurement over scales. If you look at earlier American cookbooks, the methods given are just as likely to be weighing with a scale as measuring with a cup. This is partly because many American cookbooks were in fact British—reprints of successful British recipe books such as Mrs. Rundell's *A New System of Domestic Cookery* (1807). But even in authentic American books, most of the recipes imply the existence of scales in the kitchen. The first cookbook published by an American for Americans in America was *American Cookery* by Amelia Simmons in 1796. Simmons routinely deals in pounds and ounces. Her turkey stuffing calls for a wheat loaf, "one quarter of a pound butter, one quarter of a pound salt pork, finely chopped," two eggs, and some herbs. She also gives the first American recipe for what would become a classic staple of the American kitchen: pound cake: "One pound sugar, one pound butter, one pound flour, one pound or ten eggs, rose water one gill, spices to your taste; watch it well, it will bake in a slow oven in 15 minutes."

Amelia Simmons's pound cake is not a great recipe. The very short timing, of 15 minutes, must be a typo (pound cake in my experience takes around an hour), and Simmons doesn't tell us how to mix the batter (do we add the eggs one at a time to avoid curdling? Or in one great swoop?). However flawed, it does show that in 1796 at least, Americans were not averse to putting butter and flour on a scale. Pound cake remained a favorite long after cups took over. Even Fannie Farmer includes a pound cake, not dissimilar to Simmons's, except that she had replaced the rosewater and spices with some mace and brandy; and she plausibly says that it will take one and a quarter hours in a "deep pan." And she has replaced the pounds with cups.

By the middle of the nineteenth century, cups were taking over from pounds altogether in America. At first, the cup would have  been any breakfast cup or mug that was on hand. This is how traditional cooks still do much of their measuring in countries from India to Poland. You take a glassful of this and a cupful of that and it works just fine, because you've made the dish a hundred times before, using the same glass or cup. The problem comes only if you attempt to instruct others outside your family or narrow community in how to make a certain thing, when the recipe gets lost in translation. What was different about the cup measurements of nineteenth-century America was the shift from using cups to using *the* cup—a single standard with precise volume.

Why were Americans so attached to their cups? Some have seen it as a feature of pioneer life, when those traveling west would carry makeshift kitchen utensils on the wagon, but wouldn't want to be encumbered with heavy scales. There must be some truth in this. In a far-flung frontier settlement, a local tinsmith could rustle you up a cup, whereas scales were an industrial product, made in factories and sold in towns. Besides, frontier meals tended to be ad hoc, such as johnnycake, a stodgy mess of cornmeal and pork fat thrown together from cupfuls and handfuls of this and that.

Yet the frontier mentality cannot entirely account for America's wholesale adoption of the measuring cup. The evidence from cookbooks is that measuring cups were being viewed not as an inferior substitute for scales, but as *better than them*. The cup was used in fancy well-equipped kitchens in the cities as well as in creaky wagons. Catherine Beecher, whose sister Harriet Beecher Stowe was the best-selling author of *Uncle Tom's Cabin,* wrote a cookbook—*Miss Beecher's Domestic Receipt Book,* published in 1846. Beecher notes that "it saves much trouble to have your receipt book so arranged that you can *measure instead of weighing.*" She assumes that her readers will have scales as well as cups, but views cups as handier. She advises weighing each ingredient the first time it is used, and measuring the volume of the weighed ingredient in a "small measure cup." Beecher's idea was that next time the ingredient is required, the cook will be able to bypass the scales and use only the cup.

The cup's ascendancy was helped by the kitchenware itself: the gradual emergence of specially manufactured measuring cups with gradations for half cups, quarter cups, and so on. Catherine Beecher talks of ordinary teacups and coffee cups, but in 1887, Sarah Tyson Rorer noted the recent appearance "in our market" of "a small tin kitchen cup." These cups were sold "in pairs, at various prices . . . one of the pairs is divided into quarters, and the other into thirds." This is recognizably the measuring cup as it continues to this day.

By the 1880s, it had become common for cookbook authors to give cup conversions, so that cooks could do without scales altogether. Maria Parloa, a popular cookery teacher based in Boston, gave the following conversions in 1882, using a "common kitchen cup holding half a pint":

One quart of flour	one pound
Two cupfuls of butter	one pound
One generous pint of liquid	one pound
Two cupfuls of granulated sugar	one pound
Two heaping cupfuls of powdered sugar	one pound
One pint of finely chopped meat, packed solidly	one pound

The problem with all these conversions was how to interpret them. Exactly how solid is "solidly packed" meat? How do you distinguish a "generous" from a "scant" pint of liquid? And what on earth is a "heaping cupful"?

Mrs. Lincoln, another Boston cook, and Fannie Farmer's predecessor as the head of the Boston Cooking School, attempted to weigh in with some qualifiers. A spoonful, Mrs. Lincoln noted, not altogether helpfully, was generally supposed to be "just rounded over, or convex in the same proportion as the spoon is concave."

What Fannie Farmer did was to take these measures and remove all interpretation from them. The knife with which she leveled the top of her cups eliminated all doubt, all ambiguity. Cups must not be generous or scant, heaping or packed. "A cupful is measured level. A tablespoonful is measured level. A teaspoonful is measured level." This exactitude offered the cook a sense that cookery had been elevated to the level of science.

Farmer's method was indeed a huge improvement over the heaping and scant measures of previous writers, so perhaps we can forgive her for failing to spot the fact that the entire system of cup measures was flawed.

Fannie's fixation with level measures in the kitchen reflected how late she came to cooking. She was born in Boston in 1857, one of four daughters of a printer (a fifth sister died in infancy). She never did much cooking at home. Fannie would probably have become a schoolteacher, like her three sisters, except that while still at high school, she was struck down with an illness, probably polio, and after a period of paralysis, was left permanently weakened, and with a limp. It looked for a while as if she might never leave home. In the 1880s, aged twenty-eight, she took on a job as a mother's helper in the home of a friend of the family. There she developed an interest in cooking. In 1887, she enrolled in the Boston Cooking School, one of several new schools across the country aimed at teaching middle-class women how to cook. She must have done something right, because seven years later, she was running the school, dressed in a white cap and a white apron that stretched all the way down to her ankles.

The Boston Cooking School taught Fannie Farmer to cook with the purpose-made measuring cups that had lately become available. And she in turn recognized no other method. Her entire approach was about offering cooks a sense that they could do anything, so long as they obeyed the rules and followed her instructions to the letter: absolute obedience would lead to absolute proficiency. As a latecomer to the kitchen, Farmer had none of the natural instincts to fall back on about how much of any ingredient was needed and how long to cook it. Everything had to be spelled out. She would go so far as to stipulate that the pimento garnish for a certain dish be cut three-fourths of an inch long and half an inch across.

The idea was to create recipes that would be absolutely reproducible, even if you knew nothing about cooking: recipes that "worked." She inspired the same kind of devotion as Delia Smith in Britain today. ("Say what you like about Delia," people often remark, "her recipes work.") Evidently, plenty of people found Farmer's level measures comforting, given her colossal sales (her 360,000 copies sold put her in the same league as *Uncle Tom's Cabin*, which had sold more than 300,000 in the months after publication). So long as you had your cup measures and your case knife, these were recipes you could trust, and the admirable thing about a Farmer recipe was that you could repeat it time and again with roughly the same results.

Whether we would want to achieve Farmer's results today is another matter. Her tastes have not worn well. She was fond of such things as spaghetti timbales (soggy cooked pasta reheated in a mold with salmon forcemeat) and avocados filled with oranges, with truffle decoration and a condensed milk sauce. This brings to mind the food writer Elizabeth David's comment: "What one requires to know about recipes is not so much do they work as what do they produce if they do work?"

Some of Fannie Farmer's faith in her own system came from the fact that she had rejected entirely the archaic instructions based on analogy that had constituted almost all kitchen measuring up to her lifetime. Since medieval times, recipe writers had dealt in such currency as fingerbreadths of water and butter the size of a pea, a nut, or an egg. The most universal analogy seems to have been that of the walnut. For Farmer, cup measures were superior to fingers and walnuts because they were both more accurate and more precise. In many ways she was right. Instructions such as "an egg-sized lump of butter" drive lots of rational people to despair. Today, the cookery forums on foodie websites are full of frustrated home cooks trying to ascertain the exact dimensions of a lump of dough the size of a walnut. Is it one tablespoonful? Or two?

Yet for hundreds of years, these comparisons were the main idiom of measurement in the kitchen. Here is Hannah Wolley, author of the *Queen-like Closet,* in 1672, with a recipe to make "pancakes so crisp as you may set them upright." The recipe in its entirety reads: "Make a dozen or a score of them in a Frying-pan, no bigger than a Sawcer, then boil them in Lard, and they will look as yellow as Gold, and eat very well." This is not a recipe at all in Fannie Farmer's terms. Wolley does not tell us how to make the batter, or how long to cook it.

How hot is the lard? How much of it should we use? How many pancakes do we "boil" at once? And how do we drain them?

Unless you were already very confident in the art of making pancakes, you would get nowhere with Wolley. But assuming you did have long experience in batter making and frying, it is an interesting recipe. The fancy imagery—"no bigger than a Sawcer" and "yellow as Gold"—made perfect sense if you knew your way around a kitchen. The end result—for twice-fried pancakes—sounds unusual: a bit like a pancake-doughnut hybrid; a cardiologist's nightmare, but genuinely useful for someone who wanted to make pancakes "so crisp as you may set them upright."

Before the nineteenth century, almost all recipes dealt with measurements much as Wolley did. They were aide-mémoires for those already skilled in the kitchen rather than instructions in how to cook. This is part of the reason old recipes are so hard to reconstruct: we have no idea of the quantities; we do not know the rules of the game. Take this one, from the Roman Apicius. It is for "another mashed vegetable" (the capitals are his):

COOK THE LETTUCE LEAVES WITH ONION IN SODA WATER, SQUEEZE [the water out] CHOP VERY FINE; IN THE MORTAR CRUSH PEPPER, LOVAGE, CELERY SEED, DRY MINT, ONION; ADD STOCK, OIL AND WINE.

Not to put too fine a point on it, this sounds disgusting: slimy cooked lettuce with two applications of onion, one at the beginning and one at the end. But the quantities and cooking times could make all the difference. Lovage, celery seed, and dry mint are all pungent, aniseed-y seasonings; a pinch of each might be acceptable, a whole spoonful would be horrible. Defenders of Roman cuisine say that there would have been a fine balance among all the strong flavors. We have no way of knowing if they are right.

Compared to this Apician type of recipe, which gives no quantities at all, a "piece of butter the size of a walnut" was a huge improvement.

It sounds vague, but actually it isn't, relatively speaking. Measurement is always a form of comparison—between the fixed standard and the thing being measured. In ancient societies, measuring began, naturally enough, with the dimensions of the human body. Sumerians in Mesopotamia invented units of length based on their own hands: the width of a little finger; the width of a hand; the distance from the tip of a little finger to the tip of the thumb on an outstretched hand. The basic Greek measure was the *daktylos,* the breadth of a finger. Twenty-four fingers made a cubit. The Romans took the Greek *daktylos* and made it a "digit."

Cooks in the kitchen did exactly the same thing. The finger was a measure that was ever-present. It was literally handy. "Take four fingers of marzipan," says Maestro Martino, the most renowned cook of the fifteenth century. Artusi, the Italian best-seller of the late nineteenth century, begins one of his recipes invitingly: "Take long, slender, finger-length zucchini." Using fingers to measure reflected the tactile nature of kitchen work, in which fingers were used to prod meat, form pastry, knead dough.

If there were fingers, there were also handfuls. To this day, many Irish cooks make soda bread using handfuls of flour and refuse to do it any other way. It sounds like it wouldn't work because human hands are so variable in size. But the great thing is this: an individual cook's hand never varies. The handful method may not work as an absolute measure, but it works very well on the principle of ratio.

A ratio is a fixed proportion of something relative to something else. So long as one person uses a single hand, to pick up the flour and other ingredients, the ratios are constant and the soda bread will rise. Some nutritionists today still use the human hand as a unit for measuring portion sizes: a portion of protein for an adult might be the palm of your hand (minus the fingers); for a child, the portion is the palm of a child-sized hand. In many ways, ratios work better in cooking than absolute measures, because you can adapt the recipe to the number of people you are cooking for. Michael Ruhlman recently wrote a whole cookbook founded on the principle of ratio, arguing that when you know a culinary ratio, "it's not like knowing a single recipe, it's instantly knowing a thousand." Ruhlman's ratio for bread, for example, is five parts flour to three parts water, plus yeast and salt, but this basic formula can be tweaked to become pizza, ciabatta, or sandwich bread, or it can be scaled up from one loaf to many. Unlike an Irish soda bread maker, Ruhlman constructs his ratios from precise weights, not handfuls; but the idea is the same.

Having exhausted the measuring possibilities of the human hand, cooks turned to other familiar objects. Among these, the walnut stands out for its ubiquity. The "size of a walnut" has been used by cooks as far afield as Russia and Afghanistan; England, Italy, and France; and America. The comparison has been used at least since medieval times. It has been applied to carrots, to sugar, to parmesan fritters, to cookie dough, to fried nut paste, and above all, to butter. What made the walnut so prized as a unit of measurement?

Imagine you are holding a whole, unshelled walnut in the palm of your hand, and its value becomes clearer. Like a finger, the walnut was a familiar object; almost everyone would have known what one looked like. "The size of a walnut" was much more helpful than the other common shorthand, "nut-sized" (which always begs the question: what kind of nut?). Even now, most of us could estimate the size of a walnut pretty accurately, even if we only see them once a year for Christmas. Unlike apples or pears, which come in many shapes and sizes, walnuts are relatively uniform. It is true that there are freakishly small varieties of walnut, notably the French *noix noisette*, no bigger than a hazelnut. Usually, however, when we speak of walnuts we mean *Juglans regia*, which was cultivated in ancient Greece, to which it was imported from Persia. By 400 AD, it had reached China. It was an important crop in medieval France, though it did not reach Britain until the fifteenth century. The great thing about the Persian walnut, apart from its rich oily taste and delicate brain-like kernels, is its constancy of size. Its fruits do not vary much between around 2.5 and 3.5 cm in diameter, a handy quantity. Picture a walnut on a spoon. Now imagine the walnut has transformed into butter. It's a decent amount, isn't it? A walnut is somewhat more than a smidgen and less than a dollop. Much the same as a knob.

In numerous recipes where butter is required, a walnut is indeed just right. In 1823, Mary Eaton used a piece of butter "the size of a walnut" to stew spinach. Mrs. Beeton in 1861 advised walnut-sized butter for broiling rump steaks. Fannie Farmer might protest: how do I know if my butter is really walnut-sized? But the more confident you feel in the kitchen, the less you worry about these things. "Butter the size of a walnut" reflects the agreeable fact that in most forms of cooking—baking partially excepted—a tad more or less of any ingredient is not critical.

A walnut was not always the size needed, however, and cooks developed an entire vocabulary of measurements based on other familiar objects. The analogies chosen depended on time and place. Peas were common; as was the nutmeg, to signify a quantity somewhere in the region of our teaspoon. Seventeenth-century cooks wrote of bullets and tennis balls. Coins were another useful reference point, from shillings and crowns in England to silver-dollar pancakes in America.

These measurements through analogy feel like a window onto the domestic life of the past. They reveal a shared world of imagination, in which nutmegs, bullets, coins, and tennis balls are all forms of currency. These quantities may not be "scientific"; but they reflect great consideration on the part of recipe writers, attempting to translate their dishes into terms others would understand. Elena Molokhovets was a skilled Russian cook of the late nineteenth century. Her recipes are thick with these comparative measurements. When she rolled pastry, it might be to the thickness of a finger or of two silver ruble coins. Molokhovets cut ginger the size of a thimble, and dough the size of a wild apple, and butter—what else?—the size of a walnut.

We still rely on a shared imagery of measurement. When we "dice" vegetables, we are harking back to older cooks like Robert May who cut beef marrow into "great dice" and dates into "small dice." When celebrity chef Jamie Oliver tells us how big to form the chopped meat for a homemade hamburger, he may not evoke walnuts or wild apples. But he does tell us to make it as big as a smallish cricket ball.

Quantity is just the beginning. The two hardest things to quantify in the kitchen are timing and heat. "Hold out your left hand," says the Canadian chef John Cadieux, in a voice that tells me he is used to being obeyed. We are sitting at a darkly lit table at Goodman's City, a steak restaurant in London near the Bank of England. Cadieux is the executive chef. We are talking steak. "Now take your right forefinger." He shows me how to use this finger to touch the palm of my left hand, the fleshy part at the base of the thumb. "That's what rare steak feels like," says Cadieux. My finger sinks into my own squishy flesh: it feels just like raw meat, which doesn't bounce back. "Next, bring your left forefinger and thumb together and continue to feel the base of the thumb with your right finger— that's medium rare. Add your middle finger—medium. Ring finger— medium-well done. Finally, the little finger—that's well done." I am genuinely amazed at how the flesh behind my thumb tightens with each additional finger, like steak cooking in a pan. Cadieux, a shaven-headed thirty-something who has been working in high-end steakhouses for more than seven years now, sits back and grins. "It's an old chef's trick," he says.

The restaurant has state-of-the-art charcoal ovens (two of them, $20,000 apiece), numerous digital timers lined up to cope with the endless stream of different steak orders, and the best meat thermometers money can buy. Cadieux insists on training his chefs for a minimum of two weeks (on top of whatever training they already have) before he lets them cook a steak. They must memorize the exact temperatures required for every cut and every order. Cadieux applies different standards to himself, however. "I don't like thermometers," he says. "I'm a romantic." He has cooked so many thousands of steaks, he can tell instantly by the look and touch of a steak whether it is done to order.

Which is all very well until Cadieux needs to translate his own superior knowledge for his apprentices. At that point, he gets over his dislike of thermometers. Even if he himself has no need of measuring devices, he gets his sous-chefs to use them as crutches until they develop the instinctive knowledge of a master. For the medieval master chef, this question of passing on culinary skills was much harder. He would have had the same practical knowledge of cooking as Cadieux but none of the digital probes and timers to fall back on. How did you know when a dish was done? You just knew. But this wouldn't help you to explain the principles to someone who didn't "just know." For this, you needed a range of ciphers to act as translations. Luckily, the medieval master had far longer than Cadieux's two weeks to convey the finer points of measuring to his apprentices, most of whom started work as children, watching and absorbing the secrets of timing over many years.

Cooks have always needed to measure time, one way or another. The kitchen clock, quietly ticking on the wall, is one of the least recognized but most vital pieces of technology. No one seems to know when it first got there, though it had certainly arrived by the eighteenth century. We can tell that kitchen timepieces were not the norm in medieval and early modern times from the number of recipes giving timings not in minutes but in prayers. A French medieval recipe for preserved walnuts calls for boiling them for the time it takes to say a Miserere ("Oh wash me and more from my guilt . . . "), about two minutes. The shortest measurement of time was the "Ave Maria," twenty seconds, give or take. You might say that such recipes reflect the fact that medieval France was a society in which religion permeated everything. Yet this timing-by-prayer had a very practical underpinning in an age when clocks were rare and expensive. Like the walnut-sized butter, these timings depended on communal knowledge. Because prayers were said out loud in church, everyone knew the common pace at which they were chanted. If you asked someone to "boil and stir the sauce for the time it takes to say three Pater Nosters" or "simmer the broth for three Lord's Prayers" they knew what this meant. And so far from being otherworldly, it was more sensible advice than some of the more secular examples in written recipes such as "Let the solid particles precipitate from the mixture for the time a person would take to walk two leagues." The use of prayers as timing devices belonged to the long centuries when cooks had to use deep ingenuity and vigilance to ensure that a meal came out right: cooked, but not burned.

If time was measured through prayer, heat was measured through pain. To test the heat of an oven, you reached inside. This is how bakers still proceed in many parts of rural Europe. You would put a hand in the oven and gauge from the level of pain whether the oven was ready for baking loaves—which require the fiercest heat.

One step up from this was the paper test. This was much used by confectioners in the nineteenth century. The point here was not to gauge the fiercest heat as you stoked a fire up, but the subtle gradations of gentler warmth as the oven cooled down, suitable for baking cakes and pastries, whose high butter and sugar content made them much more liable to burn than bread. Each temperature was defined by the color a sample of white kitchen paper turned when put on the oven floor. First, you had to put a piece of kitchen paper inside the oven and shut the door. If it caught fire, the oven was too hot. After ten minutes, you introduced another piece of paper. If it became charred without burning, it was still too hot. After ten minutes more, a third piece went in. If it turned dark brown, without catching fire, it was right for glazing small pastries at a high heat: this was called "dark brown paper heat." Jules Gouffé, chef at the Paris Jockey-Club, explained the other types of heat and their uses. A few degrees below dark brown paper heat was "light brown paper heat, suitable for baking vol-au-vents, hot pie crusts, timbale crusts etc." Next came "dark yellow heat," a moderate temperature, good for larger pastries. Finally, there was the gentle heat of "light yellow paper heat," which, said Gouffé, was "proper for baking manqués, génoises, meringues." A variation was the flour test, which was the same but with a handful of flour thrown on the oven floor: you were meant to count to forty seconds; if the flour slowly browned, it was right for bread.

All of this to-ing and fro-ing was done away with at a stroke once an integrated oven thermostat became commonplace in the twentieth century. The thermostat is one of those examples of a technology that feels as though it should have entered the kitchen much sooner than it did. Various thermometers were developed by scientists—including Galileo—as early as the sixteenth century, mostly for measuring air temperature. In 1724, Fahrenheit produced his temperature scale; and in 1742, Celsius produced his competing scale (from the melting point to the boiling point of ice). The kitchen is a place where plenty of water boils and ice melts, yet for hundreds of years no one thought to bring a thermometer to bear on the question of what temperature to use to bake a cake. By the 1870s, people routinely spoke of the weather in relation to ther-mometers—in 1876, English cricketers played on a "blazing July day" when "the thermometer in the sun stands at about 110 degrees," wrote the *New York Times*—yet when they stepped into the kitchen, they were still happy with "dark yellow heat" and "light yellow heat."

Finally, around the turn of the twentieth century, cooks started to see that thermometers might be rather useful after all. An American oven called the "new White House" advertised itself with an oven thermometer, included "in order to keep . . . strictly up to the minute." The first fully integrated gas oven with thermostat was mar-keted in 1915, and by the 1920s, electric stoves fitted with electro-mechanical thermostats were being produced. But the easiest thing, for those who already had a stove, was to buy a stand-alone oven thermometer and get it fitted to your existing oven.

One of the first cookbooks written after these newfangled oven thermometers came on the market was *Mrs. Rorer's New Cook Book* in 1902 by Sarah Tyson Rorer. Mrs. Rorer was the principal of the Philadelphia Cooking School, a woman with twenty years' teaching cooking. She greeted this new device with nothing less than ecstasy. Thermometers, she wrote, cost only $2.50 and "relieve one of all anxiety and guesswork." A note of early adopter pity creeps into her voice when she writes of those who are "without a thermometer" and must "guess at the heat of the oven (a most unsatisfactory way)." Rorer gave all her recipes in Fahrenheit (though she also gave a system for Celsius conversion). She clearly loved her new toy and the precision it appeared to guarantee. She thrust her thermometer into freshly baked bread and boiled meat ("plunge a thermometer into the center of the meat, and to your surprise it will not register over 170° Fahr."). Rorer liked deep-fried oysters, an authentic old Philadelphia dish. Now she could abandon the device of putting a cube of bread to sizzle in the hot fat for deep-frying, waiting to see how quickly it browned. A thermometer told her at once if the fat was hot enough. Above all, Rorer used it for measuring oven heat. With the new ability to have a thermometer installed in any kind of "modern range," whether gas-fired, coal, or wood, the cook was freed from the need to stand and watch and make "unsatisfactory attempts to ascertain the true heat of the oven." All the old worry was gone, because the responsibility for guessing what was meant by "a moderate, moderately cool or quick oven" was lifted from the harried cook's shoulders.

With the new thermometer, all that fussing was a thing of the past.

> A potato will bake in three-quarters of an hour at a temperature of 300° Fahr.; it will harden on the outside and almost burn at a temperature of 400° in twenty minutes, and if the oven is only 220° it will take one hour and a quarter to a half.

The worry was gone because, as with Fannie Farmer's cup measures, any need for individual judgment was also gone. There was no squinting at a bit of paper and wondering if it was closer to yellow or brown. You only had to follow the system and you would be fine. At least by some people's standards.

When Nathan Myhrvold investigated ovens, he found that the thermostat in "nearly all traditional ovens is just plain wrong."

The margin of error on the average thermostat is so high that it gives a false sense of security: the temperature dial in which we mistakenly place our faith is not a true reflection of what is going on inside the oven. Myhrvold calls the oven thermostat an "underwhelming" technology.

For one thing, thermostats only measure dry heat and take no account of humidity. We know that moisture content in the oven hugely affects the way something cooks: whether it roasts or steams or bakes, and, if so, how fast. Yet cooks—until Myhrvold—have hardly ever thought to measure humidity: the technical term is wet-bulb temperature. An oven thermostat cannot measure how a leg of lamb's cooking time is affected by a glass of wine sloshed in the pan; or how a bread's crust is softened by a splash of water thrown on a scorching oven floor.

That's the first problem. A greater drawback is that most domestic thermostats don't even measure dry heat very accurately. A thermostat makes its reading based on a sensor probe filled with fluid—similar to the old mercury-filled glass thermometer used by doctors in days gone by. The location of the probe can skew our impression of how hot an oven is. The probes that Myhrvold most dislikes are located "behind the oven walls," which may have a much lower temperature than the main body of the oven; it's better to have the sensor protruding inside the oven cavity, though even this is not perfect, because the further the food is from the sensor, the less likely it is to be accurate. Myhrvold found that thermostats mismeasure the dry-bulb temperature in a domestic oven by "as much as 14°C/25°F," which could make the difference between success and failure in a recipe. Every oven has its own hot spots. The answer is to calibrate your own oven: place an oven thermometer in different spots in the oven as it heats, then write down the true measurements and proceed accordingly.

In making the average home-cooked meal, temperamental ovens are a fact of life. Once you have learned that your oven gets too hot or too cold, you can turn the dial up or down, like tuning a musical instrument. This kind of rough-and-ready adjustment would not be accepted, however, in the modernist restaurant cooking of the early twentieth-first century, which sets great store by fantastically precise and accurate measuring devices. Chefs who cook in the style of Ferran Adrià of the now closed El Bulli in Spain need to be able both to measure very large amounts (up to 4 kilos) and very small amounts (to within 0.01 g) with the same degree of accuracy over and over again. Most kitchen scales, even digital ones, fall far short of these exacting standards. The solution is to have not one but two sets of scales, both of laboratory standard, one for large quantities and one for small.

Weight and temperature are by no means the only things measured in the modernist kitchen. These high-tech cooks are like explorers mapping new culinary lands. They wish to quantify everything, from the heat of chilies (measurable on the Scoville scale) to the chill in the ultra-low-temperature freezers they favor. If they wish to test how tart a fruit puree tastes, they do not use their tongue, but whip out an electronic pH meter that can give an instant and exact reading on how acidic or alkaline any fluid is. To gauge the sugar content in a sorbet mix, they use a refractometer, a tool that responds to the way light bends as it passes through a given material. It will bend more or less depending on the density of a given liquid, which in turn tells you how sweet a syrup is (sweeter is denser). This is a technological step up from the old saccharometers used by brewers and ice-cream makers from the eighteenth-century onward, consisting of a calibrated glass bulb that measured sugar content through the principle of buoyancy (the higher the bulb floated, the sweeter the substance). Before that, mead makers would drop an unshelled egg into the honey-sweet liquid; if it floated, it was sweet enough.

Chefs today are measuring things no one ever thought to measure before, such as the exact water content a potato needs to make the ideal French fry. Heston Blumenthal, visionary chef of The Fat Duck in Britain, prides himself on his Triple-Cooked Chips (cooked once in water, then in a sous-vide bath, finally in groundnut oil—I have tasted them only once and they were indeed superbly crisp).

He has found that the perfect "consistently crunchy" French fry can only be made from a potato with a dry-matter content of around 22.5 percent. "The problem," Blumenthal has noted, "is that there is no easy way to look at a potato and know how much water it contains." The answer is a special "dry-matter" scale that determines the water content of a small sample of raw potato by weighing it and cooking it simultaneously. It then finds the difference in weight between the cooked potato and the raw potato, in other words, how much water has evaporated.

Measures such as these undoubtedly help professional chefs achieve dependable results. Blumenthal knows that his Triple-Cooked Chips will always be as near the same as possible. I'm not so sure whether ultraprecision is what the average home cook is looking for, though. I glance through a Heston Blumenthal recipe for "sand," part of a recipe for something he calls "Sound of the Sea." It calls for 10 g grapeseed oil, 20 g *shirasu* (baby eels or anchovies), 2 g blue shimmer powder, 3.5 g brown carbonized vegetable powder, and 140 g "reserved miso oil," along with various other baffling things. Having measured these peculiar ingredients on our laboratory scales, we are supposed to sauté and grind them until they become a kind of savory sand. The whole recipe is deeply intimidating.

Even if I possessed brown carbonized vegetable powder—and alas, I have rummaged in my kitchen cupboards in vain—I have neither the technology nor the patience to weigh out 3.5 grams of the stuff. This is cooking as pure mathematics: everything is quantified; nothing is left to chance; there is no room for variation or judgment. For restaurant chefs who want to produce the same—often spectacular—results time and time again, Blumenthal's way makes sense. Blumenthal is the master of food as theater, and the performance only works when everything is just so. The imperatives of home cooking are different. We'd rather have flexibility than absolute control.

What if I wanted to substitute something else for the blue shimmer powder (or preferably, leave it out altogether)? What if my *shirasu* taste saltier than Blumenthal's? Pointless to ask. I have nothing to compare this recipe with, and therefore no way of knowing how it could be tweaked. Such hypermeasuring makes the average cook feel lost in a sea of numbers. Blumenthal's measures may be accurate, precise, and consistent, but no one could accuse them of being easy. Nor are they meant to be. They are aimed at chefs like him whose ambition is to push food in extraordinary new directions.

Compare and contrast with Fannie Farmer's trusty old cup measures. For all their faults—and, as we have seen, they are many—they do have one huge virtue. For cooks who have learned to cook using cups, they bestow a feeling of calm competence. They may not score highly on precision or consistency, but they are wonderfully easy. When asked to measure three level cups of flour, you think: yes, I can do this. Scoop and sweep, scoop and sweep; one, two, three. To measure with cups requires such little expertise, it can be done by a child who has only just learned to count and in a kitchen with the most minimal equipment. Because Fannie Farmer came so late to cooking, she remembered what it felt like to be perplexed in the kitchen. She herself had found reassurance in her level cup measures and warmly passed this reassurance on to her readers. Blumenthal's recipes seek to amaze, to confound, even to disgust. Farmer hoped her directions would "make many an eye twinkle." For the thousands who bought her book, reading Fannie Farmer was like having a friendly but firm red-haired woman holding your hand as you cooked, whispering: follow me and it will work.

Fannie Farmer's cup measures may not have bestowed the accuracy they promised. But she understood something every bit as important: the technology of measuring in the kitchen needs to be tailored to the person doing the measuring. Most chefs and food writers have been cooking for so long, they forget what it feels like to be thrown into a panic by the simplest of recipes.

In 2011, Tilda, a leading rice brand, conducted a UK focus group of around 500 people looking at factors inhibiting the British public from buying rice. They found that many households possessed no kitchen scales. Even when they did, there was a widespread terror of getting the measuring wrong: of overestimating the portion size or cooking the rice for too long. For many, the focus group revealed, this terror was enough to stop them from buying even the smallest half-kilo bag of rice in the supermarket: the risk of failure was too high. This stood in stark contrast to customers in Asian communities in Britain, who buy their Basmati in 20-kilo sacks from the Cash and Carry and cook it with effortless confidence, using a thumb to measure the correct quantity of water every time, just like their mothers and grandmothers before them. They hold a thumb on the base of the pot and measure rinsed rice up to the thumb joint, then rest the tip of the thumb on the rice and pour in water until it again reaches the joint. It is then easy to cook perfect fluffy rice by the absorption method. The technology being used here is sheer know-how. We all have thumbs; what we lack is the confidence to use them.

Lack of confidence also explains the existence of the most curious measuring spoon I have ever seen. Instead of tablespoons and teaspoons, it has: a dash, a pinch, a smidgen, and a drop. Those of us who feel reasonably relaxed at the stove might have assumed that you can't assign an exact quantity to a smidgen. We would be wrong. All of these terms now have technical definitions (as of the early 2000s, when this type of measuring spoon first started being manufactured). A dash = 1/8 teaspoon (0.625ml). A pinch = 1/16 teaspoon (0.313ml). A smidgen = 1/32 teaspoon (0.156ml). A drop = 1/72 teaspoon (0.069ml). Clearly, there is market out there for people who will not rest easy unless they can measure out every pinch of salt. Even if from the point of view of an experienced cook, the idea of measuring a single drop seems to be overkill.

Attitudes to measuring in the kitchen tend to be polarized. On the one hand, there are creative spirits who claim that they never weigh or measure anything. If you ask for a recipe from such a person, you will be told airily, "Oh, I never look at a cookbook"; if they do consult recipes, they happily play fast and loose with quantities. Every meal they cook is pure invention, pure instinct: cooking is an art and cannot be reduced to numbers. At the opposite end of the spectrum are those who want to assign an exact figure to everything. They view recipes as strict formulas, not to be tampered with. If a recipe calls for 325 ml double cream and a carton contains only 300 ml, then such people will anxiously buy a second carton to make up the shortfall. If a recipe says tarragon, they wouldn't dream of using chervil instead. People in this second group are more likely to think that what they are doing is scientific, the idea being that the more we can measure and pin cooking down, the more like science it will be.

Both groups are probably deluding themselves. Artistic cooks do far more measuring than they admit. And cooking-by-numbers cooks are much less scientific than they pretend.

Cooking by numbers is based on a subtle misunderstanding of the scientific method. The popular view of "science" is one of unswerving formulas and a set of final answers. In this reading, scientific cooking would be able to come up, once and for all, with the definitive formula for, say, béchamel sauce: how many grams of flour, butter, and milk, the exact temperature at which it should cook, the diameter of the pan, the precise number of seconds for which it should simmer and the number of revolutions of your whisk as it cooks: cooking by numbers. The problem with this— apart from the fact that it leaves no room to improvise, which is half the joy of cooking—is that no matter how many factors you succeed in pinning down, more spring up that you haven't thought to measure or that are beyond your control: where your flour was milled and how old it is, the ambient heat in the kitchen; whether you actually like béchamel.

Often, with all this focus on numbers, the really important thing gets overlooked. Take seasoning. It is striking how often cooks and chefs who are otherwise wedded to the numbers game do not quantify the salt content in a recipe. Nathan Myhrvold in *Modernist Cuisine* weighs everything, gram for gram, even water, yet advises that salt is "to taste." Similarly, Heston Blumenthal measures the dry-matter content in his potatoes but does not measure the salt and pepper in his signature mashed potatoes. This underscores the point that no kitchen formula can ever be definitive.

The scientific method is far more open-ended than is generally allowed. It is not a dogmatic set of numbers but a process of forming and testing conjectures based on experience using controlled experiments, which then throw up new conjectures. The process of cooking supper every night can certainly be understood in this light. My experience tells me that lemon and Parmesan taste delicious together, particularly in a pasta sauce. This leads me to form the conjecture that lime and Parmesan might go well together, too. I test for this by tossing some lime zest into tagliatelle with olive oil, basil, and parmesan one evening. We eat it. No one asks for second helpings. My provisional conclusion is that: no, lime and Parmesan do not improve one another, but further work needs to be done to eliminate the possibility that the oil was the rogue element.

Some of the wisest words ever written on the subject of weights and measures in the kitchen appear in *The Zuni Café Cookbook* by the California chef Judy Rodgers, whose approach to cooking is both very artistic—her signature dish is a bread and chicken salad made from rustic bread torn into variegated pieces—and very precise; she tells you exactly how to season the chicken and (without going so far as to use a pH measure) names the ratios in the tart vinaigrette with which it is dressed. She gently suggests that when professional cooks claim that they "never measure," it is "frankly, not entirely true":

> We may not take a tool to measure ingredients, or look at a piece of paper, but we measure with our eyes and weigh with our hands and scroll through memories of prior cooking experiences for the unwritten script for the current one.

Concrete numbers have their place in the kitchen, Rodgers insists, particularly for the inexperienced. Numbers are "points of reference," offering "at a minimum, a notion of scale and a sense of the relative scale of different ingredients, temperatures and amounts of time." The first time you make a dish you may need to follow the numbers fairly closely, which can help to "abbreviate the romantic but lengthy learning process one might characterize as 'guess, feel, botch, puzzle, try again and try to remember what you did.'" By the second or third time, the numbers are less important because you have started to trust your own senses. After all, Rodgers remarks, you do not need to measure "the exact amount of sugar or milk you add to your coffee or tea." Numbers, therefore, are crucial, but never the whole story. There is a world outside of measuring in the kitchen. Part of the scientific method is accepting that not everything is within the domain of science.

I am fond enough of my measuring devices—there's a quiet contentment in peering at that classic Pyrex measuring cup trying to see if stock for a pilaf has reached the 600 ml mark; or watching the dial swing around on a candy thermometer when making fudge; or using a tape measure to verify the diameter of biscotti dough. I even use my iPhone as a kitchen timer. Still, not everything can be reduced to measurements. Many things that matter in the kitchen are beyond measuring: how much you enjoy the company of those you dine with; the satisfaction of using up the last crust of bread before it goes moldy; the way an Italian blood orange tastes in February; the pleasure of cold cucumber soup on a hot evening; the feeling of having a hearty appetite and the means to satisfy it.

～ Egg Timer ～

Why egg timers and not carrot timers or stew timers? Because there is very little margin of error in achieving the ideal soft-boiled egg—flowing, orange yolk; set but not rubbery white. Also, because the egg is sealed in its shell, there is no way to judge it by eye: hence the long marriage between eggs and timers.

Timing boiled eggs is almost the only practical occasion in which we still use the medieval technology of the hourglass. In this digital age, most of us have on our person several items— a watch, a mobile phone—that could time a soft-boiled egg more accurately. If hourglass egg timers endure, it is surely because of their symbolic value: to watch the sands of time running out is still a powerful thing.

Recently, the entire logic of using kitchen timers has been challenged. We use timers to test for doneness. But timers can only test doneness at one remove. Time alone becomes a stand-in for temperature plus time. A soft-boiled egg becomes known as a "three-minute egg," but the minutes are only an approximation for what is going on inside the egg. Experiments by food scientists have found that the perfect milky-soft boiled egg is achieved somewhere between 141.8°F and 152.6°F. But how can we know when the egg has reached this temperature? We are back at the problem of the shell.

In the mid-1990s, a Los Angeles firm (Burton Plastics) launched Egg-Per'fect, an egg-shaped piece of plastic that goes into the water along with the eggs. Instead of measuring time, it measures temperature. There are lines on the plastic for different types of boiled egg: soft, medium, and hard. As the eggs boil, the Egg-Per'fect slowly transforms in color from red to black. The main drawback—apart from a very faint plastic odor—is its silence. You have to stand over it like a hawk. To make the Egg-Per'fect truly perfect, it would have a little sound sensor that shouted as it boiled—Soft! Medium-Soft! Hard!—leaving you free to read your newspaper and sip your coffee, while calmly awaiting the arrival of your eggs.

5

GRIND

These cooks, how they stamp and strain and grind!
GEOFFREY CHAUCER, "The Pardoner's Tale"

OST WEEKENDS, WE MAKE PANCAKES. IT TAKES A few swigs of coffee to rouse myself enough to locate the flour, milk, eggs, and butter, but after that, it's easy. Pour the ingredients into a pitcher. Blitz with handheld blender for a few seconds until lump free. That's it. The batter is ready for pouring into a hot pan. Within minutes, a pile of lacy golden-brown crepes have emerged with scarcely more effort than pouring out a bowl of Corn Flakes. To make pancakes in the Middle Ages was not so quite so easy. In the fourteenth-century advice book *Le Ménagier de Paris* (published 1393), there is a pancake recipe. It goes like this. First, get a quart-sized copper pan and melt a large quantity of salted butter. Then take eggs, some "warm white wine" (this takes the place of our milk) and "the fairest wheaten flour" and beat it all together "long enough *to weary one person or two*." Only then is the batter done.

There is a startling nonchalance in this "one person or two." It conjures up a kitchen in which there is a standing army of servants, arrayed like so many utensils. When one underling is worn out, another steps forward. Suddenly we see that the recipe is not at all like one of ours, in which the reader is the person expected to do the work. *Le Ménagier de Paris*—which roughly translates as "The Housewife of Paris"—was written in the voice of an elderly husband to a wealthy young wife, teaching her the proper way to behave. To prove her worth, a French medieval wife of this class needed to ensure that dishes were well made. But not to the extent of dirtying her own hands. She has an entire team of human eggbeaters at her disposal. As the pancakes are being fried, "all the time" another person needs to carry on "moving and beating the paste unceasingly."

This incessant beating reflects the intense urge that wealthy palates once had toward smoothness. This desire has largely abated, now that pappy white bread and spongy hamburgers are among the cheapest foods. On a fine spring day in 2011, I sat in one of the best Italian restaurants in Britain, where the main courses cost around $50. Well-heeled families were eating Sunday lunch. Many of the diners were enjoying chewy rectangles of bruschetta, anointed with olive oil and coarse salt. There were platters of crunchy green vegetables, minimally prepared. A pork chop came on a massive bone, proving a challenge even for a steak knife. Linguini with crab and chili was genuinely al dente: you felt the hard center of each strand on your teeth. Until the silky gelato for dessert, nothing was smooth; the textures were all rustic, variegated, challenging. This was not a sign of careless cooking: in the age of the food processor, it takes great conscious effort to produce a meal like this.

Until modern times, by contrast, the meals that took the most effort to make were highly processed. Popes and kings, emperors and aristocrats did not want too much to chew on. They expected fine pastes to be pounded for their pleasure in mortars with pestles. In wealthy kitchens, pastries and pastas were rolled out so fine you could look through them (by implication so fine that someone's arms ached). Sauces were strained and strained again through ever-finer sieves and cloths. Flour was "bolted" through "crees" and linen. Nuts were ground as fine as dust and made into biscuits with super-refined sugar. Now we use the word "refined" to mean rich or posh, but originally refining referred to the degree to which a food was processed. Refined food was what refined people ate.

It would be going too far to say that the only appeal of this style of food was to cause pain to the servants who made it. There were many motivations. Soft mixtures were potentially desirable in any era before modern dentistry. Medieval cooks made "mortrews," mortar-pounded concoctions of white boiled meats and almonds that were ideal for those with bad teeth. Moreover, the mingling together of many pounded ingredients corresponded to medieval ideas about temperament and balance. Later, in Renaissance times, processing became a kind of alchemy: a desire to distill things down and down and down, until all you were left with was the very soul or inner kernel of a particular foodstuff.

But when considering the technologies of grinding, pounding, and so on, we cannot get past the labor question and preindustrial patterns of work. Highly processed foods were favored by the rich not despite the labor that they caused—the number of people that they wearied—but because of it. To serve a dish such as ravioli stuffed with a pounded mass of capon breast, grated cheese, and minced herbs topped with powdered sugar and cinnamon reflected your status. Everyone who ate it knew that it would take a lot more than just a wife with a wooden spoon to rustle it up. With no electric food processor to help, such a dish required one person to knead and roll the pasta, another to cook and pound the capon, a third to grate and mince the cheese and herbs, and so on. The luxury was not just in the ingredients but the trouble it took to assemble them (as is still the case in Michelin-starred kitchens: at Ferran Adrià's El Bulli, making a rum and sugarcane cocktail required two people with handsaws to cut the tough cane into manageable pieces, another two with cleavers to strip off the bark, and two to eight more people to cut the sugarcane into sticks; all of these people were unpaid "stagiaires").

From time to time, voices have been raised against this laborious way of cooking, as much on grounds of aesthetics as anything. The Roman philosopher Seneca wrote in praise of simpler cuisine: "I like food that a household of slaves has not prepared, watching it with envy, that has not been ordered many days in advance or served up by many hands." Similarly, in the fourth century BC, there was a generation of young cooks who reacted against the ubiquity of the mortar in the Greek kitchen. In place of all the pounded mixtures of vinegar and coriander, they served up simple pieces of fish and meat, eschewing the pestle.

Despite these odd moments when pastoral simplicity became fashionable, highly refined food remained the norm on wealthy tables well into the twentieth century. Edwardians ate crustless cucumber sandwiches and drank triple-strained consommé. Behind every course of a grand dinner was a mini-army of minions with sore arms. Done by hand, grinding, pounding, beating, and sieving are among the most laborious of all kitchen tasks. The really striking thing, therefore, is how little impetus there was—until very recently—to develop labor-saving devices; and how little change there was in the basic equipment used. For thousands of years, servants and slaves— or in lesser households, wives and daughters—were stuck with the same pestles and sieves, with few innovations. This technological stagnation reflects a harsh truth. There was very little interest in attempting to save labor when the labor in question was not your own.

My mortar and pestle comes from Thailand and is fashioned from craggy black granite. I like it far better than those restrained white-china mortars, whose rough insides set my teeth on edge, like chalk on a blackboard. Its downside is that it is probably the heaviest nonelectrical utensil I own. Every time I take the mortar down from its shelf, I have a moment of mild terror that I will drop it. Which may explain why I don't actually get it down all that often. In my cooking life, it is an entirely superfluous piece of technology. I don't need it to grind flour or sugar, which come pre-ground in bags. Nor do I need it for pepper, which I can grind far more quickly and easily in a pepper grinder. Garlic is better crushed with the back of a knife on a chopping board. When I do use my mortar and pestle, it is a sign that I am feeling leisurely and want to experience a bit of kitchen aromatherapy. I might use it to pound up a pesto, relishing the sensation of crushing the waxy pine nuts against the coarse granite. Or I might crush the individual spices for curry powder (something I do about once a year in a spurt of enthusiasm before I get lazy and revert to the pre-ground type). In any case, the mortar and pestle is never necessary in a kitchen that also contains blenders and a food processor. It is a pleasure-giving device. I use it—or not—on a whim.

This is in stark contrast to the earliest crushing devices, whose basic mechanism was more or less identical to my mortar and pestle, but whose role was entirely different: to render edible that which would otherwise have been impossible to eat. It was a tool on which humans depended for survival. The earliest grinding implements go back around 20,000 years. Grinding stones enabled early populations to obtain calories from extremely unpromising foods: tough, fibrous roots and grains in husks. The process of making wild cereals digestible, through grinding and pounding, was difficult, slow, and labor-intensive. A grinder would be used, first, to remove husks or shells, and second, to remove toxins (in their natural state, acorns, for example, contain dangerous amounts of tannin, which partially leaches out when it is exposed to the air by the pestle). Third, and most important, it reduced the particle size of the food—whether nuts or grain—until it was as fine as dust: producing flour. Without grinding tools, there is no bread. The discovery of a 20,000-year-old basalt grinding stone near the Sea of Galilee alongside traces of wild barley suggested the very first experiments with some form of baking.

It would be several more thousand years, however, before stone grinding tools became common. Their use seems to have intensified during Neolithic times (10,300–4500 BC), which makes sense because this was the period when cereals began to be domesticated. Men started to settle down and deliberately plant grain, staying around in the same place long enough to harvest it. Now that their wives had settled in the same place, too, they had a ready pair of hands to perform the grinding of the grain. Several surviving ancient Egyptian figurines depict women at work, grinding cereals (probably barley) against a stone. Processing the necessary grain for that day's food became the larger part of many women's lives all over the world. Among the Lugbara people of Uganda, a woman is still typically buried with the smaller of her two grindstones. It symbolizes the fact the majority of her existence has been expended on the mindless, repetitive—but essential—action of grinding cereals for her family's nourishment.

What form did the first grinding tools take? The most basic method of crushing grain was through some type of quern: one flattish rock pounding relentlessly against another rock. Over time, better querns were devised in different shapes: saddle shaped or oval. The big breakthrough was the rotary quern, first seen in Iron Age Britain (c. 400–300 BC), a giant bagel-shaped stone atop a circular quern. Unlike the back-and-forth motion of earlier querns, the rotary quern used a circular motion to grind the cereal, which was more effective. Grain was poured through the hole in the top circle. Some kind of peg was then placed horizontally in a socket and used to turn the quern around and around. This whirling mechanism was a big improvement on the basic quern, but a large rotary quern still took two women to operate, one to feed in the grain and one to keep turning it. T. Garnett, who visited the Highlands of Scotland in 1800, watched as two women ground grain in a rotary quern, "singing some Celtic songs all the time."

Alongside querns, from ancient times, there were mortars and pestles. What came first was the receptacle: round, hard, and abrasive. The oldest mortars are as old as the oldest querns: the earliest known deep mortars have been found in the Levant and are around 20,000 years old. By the end of the Stone Age, mortars were sometimes incorporated into a house: there are giant basalt mortars set in the ground, as part of a courtyard where women or servants would have sat for hours grinding, grinding. It is easy to romanticize this way of life, but Middle Eastern burial sites indicate that using these grinding tools put a heavy strain on women's bodies: female skeletons show signs of acute arthritis, with the knees, hips, and ankles all severely worn by kneeling down and rocking back and forth to crush grain against stone.

One of the astonishing things about the mortar and pestle is how early its basic form and function were set. If you look at pictures of surviving ancient mortars and pestles, they would not look entirely out of place in a modern kitchenware shop: they might be a bit primitive, perhaps, somewhat rough around the edges; but some people like that look. The mortars and pestles surviving from Pompeii look entirely modern; in some ways, they are more sophisticated than the craggy Thai mortar on my shelf. Mortars underwent various refinements. Some were built with a lip for pouring out the finished product. Some sprouted legs, like a tripod, and some had a swelling base, forming a stand to stabilize the mortar during all the pounding. Fashions in shapes came and went. The Greeks and Romans favored a goblet shape (which became popular again in the nineteenth century) and in China they were fat and squat, whereas in the Muslim world of the Middle Ages, the most popular mortars were more cylindrical, veering toward the conical, and made from bronze, with many ornate Moorish patterns, a style that traveled to Spain.

The importance of the mortar and pestle goes beyond food. For centuries, it was the single most important utensil in the making of medicines and remains the international symbol of pharmacies. It was also used to crush pigments and tobacco. Still, its culinary use was probably the most vital one to ancient wielders of the pestle. The foods crushed in ancient mortars were diverse: in Mesopotamia, they ground everything from pistachios to dates. But its most important job was processing grains, creating the possibility of a staple food, something the hunter-gatherers had never known. Both querns and mortars were vital though bone-aching tools necessary to the task of generating enough belly-filling calories to get you through the day.

Over time, the emergence of professional millers freed most people from the need to grind their own grain. In a medieval village, the miller tended to be the most detested individual. Why? Because of the feeling of dependence he engendered, with his monopoly over the local flour. His mill—whether wind- or water-powered—was the essential tool without which no bread would be made. Instead of feeling gratitude, his customers felt resentment, and suspicion that he was overcharging for his services. As the "jolly" miller says in the nursery rhyme: "I care for nobody, no not I, if nobody cares for me."

The mortar, meanwhile, took up a permanent place in the kitchen as a pounder of mixtures. The biggest difference between querns and mortars is this: Both can grind food. But only the mortar can be used as a mixing bowl as well as a grinder. It is still used in this way to create Spanish *romesco,* a heady mixture of peppers, nuts, oil, vinegar, bread, and garlic. Such pastes had their counterpart in the cooking of the Middle Ages, when an entire genre of mortar-and-pestle cookery emerged, operated by large troops of strong-armed subordinates. It was believed that ingredients needed "tempering" to be put back into balance, and the mortar was the ideal vessel for this: in its embrace, honey tempered vinegar and wine tempered fish; foods were hammered into submission. If the noise-scape of a modern kitchen is largely electrical whirring sounds—the spinning of a washing machine, the buzz of a blender—that of a medieval kitchen was constant pounding.

This represented a continuity with the wealthy kitchens of ancient Rome. The food-processing implements that have survived from Pompeii include colanders and sieves, ladles, and mortars and pestles. The most famous Roman cookbook—by Apicius—contains a class of dishes called "mortaria," which are heady compounds specifically made in the mortar and pestle, consisting of various herbs and seasonings. This is how Apicius's mortaria are made:

> Place in the mortar mint, rue, coriander and fennel, all fresh and green and crush them fine. Lovage, pepper, honey and broth and vinegar to be added when the work is done.

Everything was pounded, pounded, pounded, until it was impossible to tell where the coriander began and the fennel ended. But not pounded by Apicius himself, and certainly not by whoever paid him. Professor Frederick Starr, one of Apicius's translators, writing in 1926—a time when many professional families were finding to their distress they could no longer afford servants, or not as many as their Victorian forbears—was very struck by the arduous quality of Apicius's food:

> The enviable Apicius cared naught for either time or labor. . . . His culinary procedures required a prodigious amount of labor and effort on the part of the cooks and their helpers. The labor item never worried any ancient employer. It was either very cheap or entirely free of charge.

I wouldn't call Apicius's slave-dependent cuisine so enviable. Nowadays, in the age of electric blenders and food processors, it is very easy to recreate an approximation of Apician mortaria. The only tricky part—other than figuring out the quantities—is the shopping. Rue and lovage are not exactly supermarket fodder, but if you are willing to search a bit, they are perfectly obtainable in plant form from a good garden center. Once you have the ingredients, mortaria can be made in seconds. Just feed everything down the chute of a processor and blitz: five, four, three, two, one, done. What you have is a greenish-brown slurry. The taste is sour-sweet, and muddy, with a rather unpleasant bitterness from the rue. It is like a much less appealing version of Italian salsa verde. Apart from food historians, it is hard to imagine anyone seeking out this strange amalgam, now that it is no trouble to make. Mortaria can never taste as good to us as they did to Apicius and the rich Romans he cooked for. They lack the seasoning of hard labor.

One of the greatest innovations of Renaissance cooking in Europe was the discovery that eggs could be used as a raising agent when baking (what the clever cooks didn't know was that it worked because the stable protein foam of beaten eggs holds the bubbles in the cake's structure as it cooks). This was the birth of the cake. Previously, cakes—insofar as they were made at all—were raised with ale barm or yeast, which gave a bready texture and yeasty taste. The discovery of beaten eggs made possible a vast array of sweet dishes in which the primary component was air. Beaten eggs were now used to make pillowy cakes with a much lighter sponge. British Elizabethan cooks also made yellow tarts from creamy beaten yolks, and white tarts from stiff whites, sweetened with sugar and cream. There was a vogue for syllabubs, frothed from wine, cream, and egg whites. Whites of eggs were also the crucial ingredient in that wonder, a "dishful of snow." These were a theatrical component in the banqueting course after a feast. To make a dishful of snow, multiple egg whites like seas of uncooked meringue were frothed with thick cream, sugar, and rosewater, then piled on a platter.

The trouble was, the culinary innovation of using masses of beaten egg whites was not matched by any technical breakthrough in equipment or techniques for beating them. The taste for fluffy egg whites was matched only by the arm-aching slog of working in a grand Renaissance kitchen. Without electricity to take the labor away, it was hard to beat whites enough to become a frothy, stable foam—which only happens when the egg's protein molecules have partly unfolded under contact with air, reforming as an air-filled lattice: stiff peaks. Wire balloon whisks—of the kind we still use today, usually made from stainless steel—were not common until the late eighteenth century. It is possible that individual households in Europe made their own versions of these whisks, though none have survived: an illustration in Bartolomeo Scappi's *Opera* (1570) looks remarkably like a metal balloon whisk. What is certain, however, is that such whisks were not common. If they had been, Renaissance cooks would have had no need to resort to the range of the much more unwieldy techniques they did when they attempted to get air into egg whites.

A 1655 recipe for "cream with snow" advises using "a bundle of reeds" rolled between your hands. The standard implement for whisking eggs until well into the nineteenth century was a bunch of stripped twigs tied together—usually birch (or less commonly, feathers). The advantage of these makeshift twig-whisks was that they could be used to impart flavorings to the cream or egg whites: recipes talk of tying the twigs together with peach branches or strips of lemon peel, giving the cream a peachy or lemony bouquet. The disadvantage was that twig whisks were extremely slow. A recipe from 1654 by Joseph Cooper—"chiefe cook to the late king"— stipulates "half an hour or more" for the beating of eggs for pancakes. As late as 1823, Mary Eaton, a cookery writer, advised that the egg whites for a large cake would take *three hours* to beat adequately.

When I was young, my mother used to bake old-fashioned British tea-time cakes—Madeira or cherry or Dundee—always preparing the batter with a wooden spoon in a homely ceramic mixing bowl. We creamed the butter and sugar by hand before adding the eggs. I can still remember the dull throb in my arms, the feeling of sheer enervation that came upon us by the time the butter and the sugar were fully creamed. Yet the whole process took no longer than ten minutes, perhaps a little less if we remembered to get the butter out of the fridge before we started. The degree of tiredness that must descend after three hours of whisking egg whites with nothing but some twigs simply does not compute. These were surely recipes to weary one person, or two, or three. To add insult to injury, due to superstition, it was often insisted that a cream or egg whites be beaten *all in one direction,* as if changing direction would break the spell and stop the mixture from frothing. Perhaps this superstition arose from the sheer difficulty of achieving a froth with such limited machinery—worries were also expressed that egg whites would become bewitched and not get stiff enough on "damp days."

Still, a "birchen rod" was a preferable option to some of the other technologies available. Once forks were in common use—from the late seventeenth century onward—they were at least an option. Until then, many cooks might make do with a spoon or a broad-bladed knife, neither of which offered much traction. The nastiest idea was to take the egg whites and wring them repeatedly through a sponge, a method both ineffectual and rather disgusting, particularly if the sponge had already been used for some other purpose.

No wonder chocolate mills, or moliquets, were greeted with enthusiasm when they first arrived in Britain at the end of the seventeenth century. These wooden instruments—which are still used in Mexico and Spain for foaming hot chocolate—consist of a long handle and a notched head, a bit like a water mill. They work by being spun between the palms of the hands. In the late seventeenth century, they start to show up in the inventories of large country-house kitchens, doubtless being used to whip eggs as much as to froth the newly fashionable drink of chocolate. As late as 1847, in an American cookbook, the moliquet, or chocolate mill, was being mentioned as an alternative to the birchen rod, for whipping cream. Even using the moliquet was a relatively laborious way to beat eggs.

Then again, laborious preparation was not limited to egg whites. Most recipes involving beaten whites also included sugar, or double-refined sugar to be precise, another arm-sapping thing to produce. It is easy to forget what a revolution it was when sugar began to be sold pre-ground in the late nineteenth century; when British customers could choose among caster, granulated, and icing sugar without having to do any of the work themselves. Pre-ground sugar is a far more labor-saving invention than sliced bread. Traditionally, sugar came in a lump or loaf, conical blocks ranging in size from five to forty pounds. It was "nipped" into smaller pieces using sugar nippers. In order to convert it into something to be used in cooking, it needed to be pounded—once more in the trusty mortar—and refined through a series of ever-finer sieves. Colanders and sieves are another of those implements, like the mortar and pestle, that have not really altered in essentials since ancient times, the reason being that ancient cooks relied on them far more than we do.

As late as 1874, the Paris-based chef Jules Gouffé described what was involved in processing sugar. This was how he made granulated sugar (used for sprinkling on sweet pastries):

> Procure three sifters or colanders, one with holes $\frac{3}{8}$ inch in diameter, another with holes $\frac{1}{4}$ inch in diameter, a third with holes $\frac{1}{8}$ inch, and a hair sieve.
>
> Chop the sugar into pieces with a knife, and break up each piece with the end of a rolling pin, being careful not to grind any of the sugar to powder as this would take away the brightness of the remainder.

The sugar then has to be sifted successively through each of the sifters, concluding with the hair sieve.

Gouffé complains that some do not take the trouble of going through this whole rigmarole, "owing to its being rather . . . troublesome." Instead, they merely pounded the sugar in a mortar without any of the sieving. Gouffé regretted such laziness, noting that mortar sugar lacked the brightness of sugar sifted "the old-fashioned way." He is by implication regretting the existence of kitchens less well-equipped with labor than the royal ones in which he worked. Surprisingly little had changed in the nearly 500 years since *Le*

Ménagier de Paris; or in the 2,000 years since Apicius, come to that. Sifting and grinding, beating and straining; these were still activities in which people were employed to weary themselves, in order that rich people might enjoy fluffy creams, powdery sugar, and other rich compounds.

The technological conservatism of food processing cannot be understood apart from the servant question. We often ignore the great and disquieting fact that premodern cookery books were largely written for people who did not themselves do any of the hands-on cooking, for people who would take the credit for what was served at their table without having put in any of the elbow grease. Well-born ladies might dress a salad with their own fair hands, or perform some of the prettier tasks, like sugar work, but they did not need to do any of the heavy beating and grinding tasks, because they had people to do that for them. Robot Coupe is a twentieth-century French brand of food processing machines: choppers, grinders, kneaders, and sievers. The name implies that these are kitchen robots, like an artificially powered servant. But for as long as there were plenty of real servants on hand—or a hard-working wife in poorer households—there was no call for robots.

Things only really started to change after the Industrial Revolution, when altered patterns of labor combined with factories that could mass-produce low-cost metal gadgets finally led to an outpouring of new machines designed to make the cook's life easier.

The phrase "labor saving" is first recorded in relation to manufacturing in 1791. It would be another half a century before the concept arrived in the kitchen. In the second half of the nineteenth century in the United States, the market was suddenly flooded with "labor-saving" culinary devices, many cheaply made of tin. There were raisin seeders, potato mashers, coffee mills, cherry pitters, and apple corers. Many were heavy apparatus clamped to the table like meat-grinding machines, which also appeared in volume. And all of a sudden, there were hundreds and hundreds of competing varieties of eggbeaters. What tulips were to Holland in the 1630s and Internet startups were to Seattle in the 1990s, eggbeaters were to the East Coast of the United States in the 1870s, 1880s, and 1890s. Between 1856 and 1920, no fewer than 692 separate patents were granted for eggbeaters. In 1856, one eggbeater patent was issued; in 1857, two; in 1858, three. In 1866, the number had jumped to eighteen, with designs ranging from jar shakers to tin shakers, from ratchets to the Archimedes (a kind of up-and-down mixer based on the Archimedes screw used in shipbuilding).

Marion Harland, a cookery writer who lived through the egg-beater bubble, recalled how unsatisfactory many of the novel beaters turned out to be. She found that very few of the patent eggbeaters outlived the initial excitement. Wooden handles fell off; tin handles stained your hands black. Elaborate machines consisting of "whirligigs" inside a tin cylinder seemed marvelous until you found that the tin cylinder was impossible to wash and too big to whip small quantities. "After a few trials," Harland added, "the cook tossed the 'bothering thing' into a dark corner of the closet, and improvised a better beater out of two silver forks, held dexterously together."

One of the first patent eggbeaters to outlast the initial novelty stage was the Williams eggbeater, patented on May 31, 1870, and better known as the Dover. The Dover is an American icon: the basic form that the cheapest hand-operated eggbeater will still take in any hardware store. The idea behind it is simple: two whisks are better than one. The earliest 1870 Dover beaters consisted of two bulbous beaters with a rotary wheel to turn them. Turner Williams of Providence, Rhode Island, the inventor who first thought it up, described the advantages of his invention as being the "very peculiar shearing action" that came from two wheels revolving in opposite directions at once in the same space, something not seen in any previous beater.

The Dover was an instant hit, so much so that "Dover" became the generic term for eggbeater in America. "Look for

'DOVER' on the handle," ordered an ad of 1891, indicating its huge popularity; "NONE OTHERS ARE GENUINE." An 1883 book of *Practical Housekeeping* praised the Dover as "the best in the market." The writer Marion Harland was another fan. Writing in 1875, five years after it came on the market, she insisted that "egg whipping ceased to be a bugbear to me" from the day she bought a Dover, adding that she would not sell it for $100 (note: a portable eggbeater at this time would have cost no more than 10–25 cents). What was so great about the Dover?

> Light, portable, rapid, easy and comparatively noiseless, my pet implement works like a benevolent brownie. With it I turn out a meringue in five minutes without staying my song or talk.

Harland, whose real name was Mrs. Mary Virginia Terhune, offers some insights into the circumstances, both social and culinary, that created America's eggbeater boom. Born in 1830, she was brought up in rural Virginia, the third of nine children. Her own mother did little or no cooking. "I doubt if she ever swept a room, or roasted a piece of meat, in her life," Harland later wrote. As was traditional for Southern ladies, she had "black mammies" to froth her eggs for her (and do much else besides).

Harland herself took a more active role in the kitchen than her mother. As well as producing twenty-five novels, she believed it was her calling to master the role of "homemaker." After she married a Presbyterian minister in 1856 and moved to New Jersey, Marion decided to teach herself and her cook greater skills in the kitchen. In 1873, she published the results of their many cooking sessions in *Common Sense in the Household*, which sold 100,000 copies.

Harland is not writing for women who have to beat all their own eggs. She assumes her readers will have a cook, but one who needs considerable guidance and help to produce eggs as fluffy as is desirable. Harland's ecstasy over the Dover eggbeater belongs to an uneasy transition point in the history of servants. The middle-class

American women she is writing for still have cooks, but probably only one. If the cook's arms become weary, then their arms will be next. Harland writes—in terms so condescending they make us cringe today—about the conversation she has with her servant, Katey, on bringing home an expensive fixed eggbeater, a "time-and-muscle saver in a box." Harland brings the "cumbrous" eggbeater into the kitchen, "trembling" with excitement. "Yes, mem, what might it be, mem?" asks Katey. The complicated contraption goes wrong; it ends up tipping a bowl of ten yolks onto the floor. The hapless Katey is forced to experiment with a number of other devices before Harland discovers the miracle Dover beater.

Why did fluffy eggs matter so much, anyway? The great eggbeater boom coincided with a period of American cuisine when sweet things at respectable tables had become intensely aerated. For dessert, there might be apple snow or orange snow or lemon snow, each requiring the whites of four eggs, whipped to a "standing froth." There was Orleans Cake (six eggs beaten light and the yolks strained) and Mont Blanc Cake (the whites of six eggs, very stiff). There were creams and charlottes, syllabubs and trifles, whipped frostings, muffins and waffles, not to mention meringues. All of these dishes depended on highly aerated eggs: yolks beaten to a cream, whites to a fluff. On the successful rise of these delicacies, a housewife's reputation might depend. A properly airy cake—produced with a Dover or one of the other newfangled whisks—spoke well of a household. Even though it is mostly her cook who actually does most of the whipping, Harland takes the credit for the light muffins emanating from her kitchen. She contrasts her muffins with those of a less vigilant friend who wasn't aware that her cook, Chloe, had been lazily making muffins with eggs barely beaten at all with "half a dozen strokes of the wooden spoon." Harland rebukes her friend for not having been more "alert."

The eggbeater boom answered a desire in middle-class American women not just to get more out air into their eggs, but to get more labor out of their servants. To those who were without servants altogether, the eggbeaters were supposed to make them not regret the lack; to feel that their arms were doing no work even when they actually were. In 1901, a Holt-Lyon eggbeater along Dover lines promoted itself with the claim that owing to its unique "flared dashers" that "instantly tear the eggs into the minutest particles," it could beat "eggs lighter and stiffer than the best hand whips in one-fourth the time."

Yet despite the euphoric claims of the advertisers, none of these mechanical eggbeaters was really very labor-saving. The great drawback of the rotary beater is that it requires two hands to operate, leaving you no way of holding the mixing bowl. The paddles have a tendency to jam stubbornly in a particular place as they rotate, or else to whirr around too fast. They slip around in the bowl, spattering eggs everywhere long before they have stiffened. The Dover claimed it could beat the whites of two eggs in ten seconds: nonsense. A rotary beater usually takes longer to get whites stiff than a balloon whisk, in my experience; in either case, it is a matter of minutes, not seconds.

Many subsequent eggbeater designs tried to address the inadequacies of the Dover, only to create new glitches. Various eggbeaters dealt with the problem of a sliding mixing bowl by setting the paddles into an attached jar or bowl; but this created its own annoyances, in that you could only beat a small amount at a time and the bowl attachment was another thing to wash. Other beaters addressed the problem of needing two hands to hold the beater. "A New Idea in Egg Beaters" boasted a 1902 ad for a Roberts eggbeater, a type of Archimedes whisk. This was "the only automatic beater made that works with one hand. . . . Simply press on the handle and release." This was undoubtedly an advantage; but the one-handed beaters—whose mechanisms ranged from strange wire whirls to springs to disks like potato ricers—were far from perfect. They took a very long time to froth eggs or cream, and they could malfunction if a hasty housewife attempted to speed things up. "Do not operate too fast" warned the Simplex one-handed mixer, unhelpfully. Strangest of all was a family of water-operated egg whisks hooked up to the new running water that was appearing in American homes. "Turn the Faucet and it Starts!" boasted The World Beater.

Looking back on this curious moment in American history—the eggbeater bubble—we are faced with a conundrum. From a purely technological point of view, not a single one of the hundreds of patented designs on which so much intelligence was lavished and so many dollars spent was actually an improvement in efficiency or ergonomics over the basic French balloon whisks, which had been in use at least since the eighteenth century, long before the eggbeater boom started (and possibly as long ago as 1570 in Italy, as mentioned above). No top chef now would dream of using a Dover eggbeater. But many of them still have a wide range of old-style balloon whisks (or "French whips"), sometimes using them in tandem with old-fashioned copper bowls. The top quality balloon whisks now come with insulated handles, and the wires are made of stainless steel instead of tin. Other than this, however, these are exactly the same whisks that would have been used by an eighteenth-century confectioner.

So what was going on with the American eggbeater boom? The whole thing was a phantom. This wasn't really about saving labor, because the French whip took less arm action to do the job than most of the new patented designs. It was more about the illusion of saving labor and time. Rather than offering a real cure for weariness, they were placebos. Those who bought them—like Marion Harland— needed to feel that someone, if only the manufacturer, was on their side in the perennial battle to produce the fluffiest eggs in the least amount of time. What the eggbeater boom tells us is that suddenly, cooks had started to rebel against their tired arms. But those arms would only really get a rest with the advent of the electric mixer.

If Carl Sontheimer had not been so taken with quenelles, the history of American home cooking over the past forty years might have been very different. In 1971, Sontheimer was a fifty-seven-year-old engineer and inventor (whose discoveries included a lunar direction finder used by the National Aeronautics and

Space Administration [NASA]), a graduate of the Massachusetts Institute of Technology, and a French food buff. Having successfully started and sold two electronics companies, Sontheimer was enjoying early retirement. As much a hobby as a business, he traveled to France with his wife, Shirley, looking for French culinary products that might translate in an American market. It was at a French cookery show that he spotted it: a food processor designed for restaurant use, called the Robot Coupe. It wasn't pretty or compact, but it was amazingly versatile. As well as being able to blend— like the electric blenders that had been on sale in America since the 1920s—it could grind, chop, dice, slice, and grate. It could take almost any food and reduce it to puree. Carl Sontheimer looked at this bulky machine and what he saw was quenelles.

"A quenelle," writes Julia Child, "for those who are not familiar with this delicate triumph of French cooking, is *pâte à choux*, cream, and purée of raw fish, veal, or chicken that is formed into ovals or cylinders and poached in a seasoned liquid." Prepared in the traditional manner, they are a pain to make. Soufflés are child's play by comparison. The quenelle mixture—the paste of chicken or fish— required long pounding and sieving to ensure it was satin smooth. Even Julia Child in 1961, that "servantless cook," needed to go to the trouble of putting her fish quenelle paste through the grinder twice. And this was all before you got to the perilous task of molding the fragile mixture into oval shapes using two dessert spoons. Julia Child noted, with typical kindness, that "in case of disaster," if the quenelles collapsed you could "declare it to be a mousse."

Carl Sontheimer spotted that the wondrous machine in front of him could take this fraught process and hugely simplify it. All of the pounding and sieving could be done at the touch of a button. The Robot Coupe had been created in 1963 by Pierre Verdun, a French inventor who aimed his device at the restaurant business. It was a hefty drum with a rotating blade inside. It had three functions: start, stop, and pulse. Sontheimer saw that a scaled-down version could work every bit as well in home kitchens. As soon as he had spotted the machine, he negotiated distribution rights to sell his own adapted version of the Robot Coupe in the United States. He brought a dozen machines home for his own personal experiments in the kitchen. In his garage, he fiddled with the various versions, taking more than a year to analyze every component until he created a model that made the smoothest quenelles with the greatest of ease. What to call this new wonder gadget? "He always thought of French cooking as an art and he wanted it to be based on the word *cuisine*," recalled his wife, Shirley. Hence, Cuisinart.

When the Cuisinart launched in the United States in 1973, it was expensive. The initial retail price was $160. In today's money, that is nearly $800 (based on the Consumer Price Index; by contrast, in January 2011 you could buy a brand new Cuisinart for $100). At this price, it might have been expected that the Cuisinart would never be anything other than a niche piece of equipment, and sure enough, for the first few months, sales were sluggish. All it took, however, was a couple of favorable reviews—one in *Gourmet* magazine, one in the *New York Times*, and suddenly the Cuisinart was flying off the shelves. Craig Claiborne, the food critic for the *New York Times,* was an early adopter of this "most dextrous and versatile of food gadgets." As an invention, he compared it to "the printing press, cotton gin, steamboat, paper clips, Kleenex," the equivalent of "an electric blender, electric mixer, meat grinder, food sieve, potato ricer and chef's knife rolled into one." It was, he excitedly claimed, the greatest food invention since toothpicks.

There was similar excitement in Britain, where another version of Verdun's invention was marketed under the Magimix brand, also in 1973. A writer from the *London Times* described how it revolutionized the shredding of cucumbers and carrots and made it possible to cater for an entire wedding party and still have time left over to get ready before the guests arrived.

By 1976, the cost of the Cuisinart in the States had actually gone up to $190, but even at this price, hardware stores could not always get enough stock to meet demand. At this time, Shirley

Collins was the proprietor of Sur la Table (founded 1971), which is now the second-largest US cookware retailer after Williams-Sonoma, but was then a single shop in the Pike Place Farmers Market in Seattle. A small coffee shop called Starbucks had opened just down the way. Pike Place Market sold the best fresh produce of the Seattle area; berries in the autumn, Blue Lake string beans in the summer. Collins would tailor her goods to fit the seasonal calendar. When fat green stalks of asparagus arrived in the spring, she would sell "large quantities of asparagus cookers." Collins was also the first person in the whole of the Northwest to sell the Cuisinart processor. At first she sold an "average of one a day." Soon, sales escalated, dramatically.

What Collins observed happening with the Cuisinart was something remarkable. The people who bought the Cuisinart were not like other customers who might buy a single asparagus steamer and never return. The customer who bought the Cuisinart kept coming back for more utensils: for "balloon whisks and copper pots and then for whatever was needed once they were launched on a new cooking venture." The machine had got them hooked on the whole process of ambitious cooking. It wasn't just that Cuisinart made a range of tasks easier in the kitchen "for anyone who cared to chop or slice mushrooms, make quenelles, dough or filling." Collins had observed something altogether more significant. It was a "real explosion in cooking." A single machine had transformed how many people felt about spending time in the kitchen. It was no longer a place of drudgery—a site of weary arms and downtrodden housewives. It was a place where you made delicious things happen at the flick of a switch. And $190 was not so much to pay for the transformation of cooking from pain to pleasure.

Cuisinart was by no means the first electrical mixing device on the market. Blenders or liquidizers had been around since 1922, when Stephen J. Poplawski, a Polish American, designed a drinks mixer for the Arnold Electric Company. Its original use was making malted milk shakes at soda fountains. Then in 1937 came the Waring

Blender. It was based on an earlier model called the "Miracle Mixer," which had suffered unfortunate teething problems with the seal of the jar: when it was switched on, malted milk was apt to spurt out all over the countertop. The Waring Blender worked better, and thanks to being promoted by the popular singer and bandleader Fred Waring, it was an instant hit. By 1954, the Waring had sold a million. Most electric blenders work in the same way. There is a motor underneath, a glass jar with a handle on top, and some small rotating metal blades to connect the two. Crucially, a rubber washer must be installed to prevent any smoothie or milkshake dripping on the motor. The blender is an awesome gadget. In goes fibrous pineapple, pulpy banana, lime juice, hard ice cubes, and some scraggy mint leaves. Blitz like crazy and out comes an aerated silky-smooth liquid, with the kind of consistency a Victorian servant would have needed three different sieves to achieve.

Yet the blender has its limits. Washing the glass container is one. The limited size of most domestic blenders is another. Every time I attempt to make a smooth green watercress soup in my blender, it seems to turn into one of those math puzzles that ask you to pour different liquids into different containers. You ladle half the soup into the blender and puree it. But now how do you puree the second half? You need a third container to hold the batches as they are done. Both these problems—the tedious cleanup and the limited size—were solved in one fell swoop by the immersion, or stick blender, patented as the Bamix in Switzerland in 1950, but not generally used in the home in the United Kingdom or the United States before the late 1980s. I consider this one of the greatest kitchen tools: it was a virtuoso piece of lateral thinking to bring the blender to the pot rather than the other way around. My handheld blender gets used most days, whether for emulsifying a vinaigrette, making banana smoothies, creating a ginger-garlic puree for Indian cooking, or rendering a tomato-butter pasta sauce silky smooth.

It is a marvel. Still, there are some jobs it cannot do. "Will It Blend?" is the hugely successful ad campaign produced by Blendtec from 2006 onward, in which we see a white-coated Tom Dixon, the founder of Blendtec, attempting to blend a bizarre selection of items: golf balls, marbles, an entire chicken, bones and all, mixed with Coca-cola, even an iPhone. The implication of the ads is that a blender can do anything. But it can't, even a third-generation power blender like the Blendtec (or its rival brand, the Vitamix). Blenders can grind nuts; but they can't chop meat. They can whiz a raw carrot so fast that the friction warms it into a kind of soup; but they can't shred carrot into a fine salad the way a food processor can because no matter how powerful the motor, the blades are too small.

The heavier end of food processing was met by a series of massive electric food mixers. The first to reach the market was an electric stand mixer, invented by Herbert Johnston in 1908 for the Hobart Manufacturing Company, a firm specializing in motorized meat grinders. Johnston was watching a baker struggling to mix bread dough with a metal spoon. It struck him as absurd. Surely the task could be performed more easily using a motor. The first Hobart electric mixers were industrial, with a dimension of eighty quarts. In 1919, however, Hobart launched an offshoot, KitchenAid, providing a scaled-down countertop version for restaurants, weighing sixty-nine pounds. This was then scaled down still further for the home kitchen. The KitchenAid is still the all-American mixer par excellence, a great chunk of metal, like a Humvee, but in pretty colors like a Cadillac (meringue, red, pearl gray); such a mixer turns the airy layer cakes and frostings so difficult to achieve with a rotary beater into a cinch.

The British equivalent was the Kenwood mixer, launched in 1950. It was the invention of Ken Wood (1916–1997), an electrical engineer fresh out of the Royal Air Force. Before the war, he had run his own business selling and repairing radios and televisions. Wood surveyed the existing gadgets on the market worldwide and attempted to combine the best elements in a single machine, the Kenwood Chef. Wood took a can opener from the United States, a potato peeler from Germany, a spaghetti maker from Italy and put them together with a mincer, a beater, a juicer, a liquidizer, and more. This wonder machine was capable—if you bought all the attachments—of whisking, kneading, liquefying, extracting, mincing, grinding, peeling, and also opening cans and even extruding different pasta shapes (the last function seems like showing off). It was advertised with the slogan: "Your Servant, Madam!" bringing home the point that food mixers did the work once done by human arms.

The Kenwood was and is a formidable piece of engineering. And yet the Cuisinart/Magimix was more significant and life changing. The Kenwood is all about the attachments, whereas with the Cuisinart, all you really needed was the basic S-blades it came with: the sharp, double-bladed stainless steel knife whirling like a dervish inside the plastic bowl. It was these blades that enabled the food processor not just to liquefy and mix but to chop and to pulverize. It was these blades that were revolutionary, for the first time making many cooks feel free rather than enslaved. Roy Andries de Groot was the author of one of the earliest specialist food processor books that appeared in the years immediately following 1973. The processor, he wrote, was "virtually the equivalent of having, as your constant kitchen helper, a skilled chef armed with two super-sharp chef's knives and a cutting board." What was more, it could "produce all the results achieved by a stone mortar and pestle. It can tenderize tough ingredients by slashing and reslashing their fibres, just as if they were pounded for an hour in a pestle."

The metal S-blades were not the only attachment to come with the original Cuisinart. There was also a medium serrated slicing disk, ideal for slicing raw vegetables such as carrots, cucumber, or cabbage ("Before you can say 'coleslaw,'" wrote de Groot, "the work bowl will be full of perfectly shredded cabbage"). Various grating disks could be used to julienne a cucumber or to reduce a knobbly celeriac to that classic French starter, céléri rémoulade. The most mystifying attachment was the noncutting plastic blade that came free with the machine, exactly the same size and shape as the steel blades but without the cutting power. De Groot recalled the comment of a chef who observed: "Their only use is to keep you awake at night, wondering what they are for." It didn't matter. To get the most out of a 1950s Kenwood, you needed a big range of accessories, many of which were actually as bulky to store as the gadgets they were supposed to replace (the liquidizer attachment was practically as big as a stand-alone blender). With a food processor, the attachments were more compact and you needed to use them less. Almost everything could be done with those basic metal S-blades, whirling in the bowl as you fed ingredients down the plastic chute.

They could be used for chopping hamburger meat and whizzing cake batters; for mincing onions and making the easiest mayonnaise in the world. Nearly forty years after the first Cuisinart, the food writer Mark Bittman was still marveling at this feature of the food processor:

> By-hand instructions for mayo require you to dribble oil—not quite drop by drop, but close—into an egg-acid mixture, while beating with a fork or whisk. It's doable and it's fun—once.
>
> By machine, you put an egg, a tablespoon of vinegar, two tea-spoons of mustard and some salt and pepper into a bowl; you put the top on and start 'er up; pour a cup of oil into the pusher, with its little hole, and go sip coffee or do yoga. The oil drizzles in, and you get perfect mayonnaise in a minute. That alone is worth the price of admission.

The food mixers of the mid-twentieth century made it easier for housewives to do many of the jobs they were doing anyway: mincing meat, beating eggs, stirring cake batter. The food processor took it a step further, encouraging its owners to embark on types of cooking they would once have deemed impossible.

In 1983, the British cook Michael Barry noted that in the past "only a few brave and dedicated souls ever tried to make pâté at home" because of the "exhausting process of cutting up, mincing, blending and then cleaning up the equipment." Now, making pâté had become normal, the work of five minutes: "The processor has changed our way of life." At a stroke, the processor demystified many of the trickiest dishes in the repertoire of French haute cuisine, including Carl Sontheimer's beloved quenelles. Once, the wealthy of Europe wearied their servants to enjoy these puffy morsels. No more. To make the mixture, now you could simply fling two boneless chicken breasts, salt, pepper, parmesan, cream, and egg into the bowl of the processor and press the button.

So great is the feeling of freedom that the food processor brings to its middle-class devotees—I include myself—we should be careful not to delude ourselves that it has really saved all labor. The medieval housewife making pancakes in *Le Ménagier de Paris* stood face to face with the people she was wearying, whereas our servants have mainly been removed from view. We do not see the hands in the chicken factory that boned the breasts, never mind the chickens that gave their lives, nor the workers who labored to assemble the parts of our whizzy food processors. We only see a pile of ingredients and a machine ready to do our bidding. Alone in our kitchens, we feel entirely emancipated.

Every revolution has its counterrevolution. You can't unleash something as extraordinary as the food processor on the world without a backlash. In the case of the Magimix in Britain, the backlash came early. In 1973, the year it arrived on the market, a writer in the *London Times* suggested it would deprive future generations of the pleasures of podding peas and kneading dough by hand. She even went so far as to suggest that by depriving cooks of tactile stimulation, the processor might leave us all in need of "group therapy."

No one could displace the processor from our lives once it had arrived. But they could complain, always in the same terms. It took the joy out of cooking; it would result in robotic food, which couldn't possibly taste as good as something handmade and artisanal; and it reduced everything to mush.

In fairness, there was something in this last complaint. The birth of a new gadget often gives rise to zealous overuse, until the novelty wears off. Reading through the early food processor cookbooks of the 1970s and 1980s, it is striking how many of the recipes had the consistency of baby food. Any vegetable that could be pureed, was. There were endless pâtés and timbales, dips galore (taramasalata, houmous, baba ganoush), and strange concoctions in ring molds. In those early years, many restaurants as well as home cooks couldn't help themselves from blitzing everything in their new toy. Quenelles went from being a rare aristocratic treat to a weeknight supper dish, until eventually people discovered that with the rarity factor gone, they were not so special after all. When was the last time you ate a quenelle?

In 1983, the food writer Elizabeth David noted the connection between the widespread availability of the food processor and nouvelle cuisine, with its obsession with velvety purees. She was dining in a "much praised" London restaurant with Julia Child in the 1970s when the latter noted that what they were eating was "Cuisinart cooking":

> About seven dishes out of ten on that restaurant menu could not have been created without the food processor. The light purees, the fluffy sauces and the fish mousselines so loved by today's restaurateurs can also be achieved at home more or less by pressing a button. . . . [I]t is indeed a marvel that the food processor does all the mincing, chopping, pureeing and blending, without a thought of all that hard pounding of the past. But let's not treat the food processor as though it were a waste-disposal unit.

Thanks to David herself, among others, the pendulum of food fashion swung back again, to more robust French and Italian provincial cooking, in which the individual ingredients are discernible. Soups and stews became chunky, which was a way of parading that no food processor had been used in their preparation. Fine-textured food lost almost all its previous cachet. Now it was the rustic and the irregular that was prized, because this showed that someone's hand had been tired out making it.

The mortar and pestle came back into fashion. Food writers would insist bossily that a pesto, a Thai curry paste, or a Spanish *romesco* could only be made authentically in a pestle. The processor kind could never *possibly* taste as good. There was even nostalgia for the way of life of all those women in Italy, Spain, Africa, or the Middle East who sat around communally pounding the day's food for hours on end, singing all the while. It didn't seem to occur to these food writers that perhaps the women were singing because it was the only way to stop their mouths from screaming with boredom at their task. While we in the cities of the West were busily imitating the old peasant ways, many of the peasants had switched to using food processors. In 2000, the California food expert Marlena Spieler traveled to Liguria to research how pesto was made in its birthplace. What she found was that "after proudly showing off the huge, heirloom, generations-old mortar and pestle, most Ligurians will then show you what they really use for making pesto: a food processor."

The same was true in the Middle East. By 1977, more food processors were in use per capita there than anywhere else in the world. One reason for this was *kibbé*, a dish that takes many forms, both raw and cooked, all of them involving finely pounded lamb, usually with bulgur wheat, cinnamon and allspice, onion, and green herbs. The Lebanese writer Anissa Helou remembers *kibbé* being made at home in Beirut by her mother and grandmother:

> They sat on low stools, either side of a beautiful white marble mortar in which were pieces of tender pink lamb. The rhythmic sound of pounding swelled from a slow dull beat to a faster, louder one as the meat was pounded into a smooth paste.

The pounding process took an hour, during which Helou and her sisters "darted in and out of the kitchen," asking if it was ready yet. Next, the pounded meat had to be made into "well-formed balls" with the bulgur and seasonings. This stage must still be done by hand. But the pounding—which previously took two educated women an hour to perform—now takes a minute's pulsing in a machine.

This is exhilarating; but it is also a little bit slighting, to the skillful hands that had pounded *kibbé* for so many generations. It is what happens whenever a machine replaces the labor of an artisan: the artisan's skills become devalued. The food processor was an affront to the proud ego of a cook, because it made the effort superfluous. All that pounding could feel worthwhile if you told yourself that your hands and your hands alone were what made the difference between good *kibbé* and indifferent *kibbé*. By doing the same job just as well, if not better, the food processor stripped the hardworking cook of some of her dignity. By working so well, this machine seems to discount the effort it once took to process different foods: to whisk a mayonnaise, sieve a smooth puree of carrots, pound *kibbé*.

The Thermomix is a newish device that makes the cook's hands more or less irrelevant. It is advertised as being more than ten kitchen appliances in one. This is a blender and processor that can also weigh, steam, cook, knead bread dough, crush ice, emulsify, mill, grate, and puree. The Thermomix can do many of the subtle tasks for which human hands were once considered most necessary. After you have plonked in the ingredients, it can stir and cook a creamy risotto. It can make velvety lemon curd and perfectly emulsified hollandaise. Your only task is to eat the results.

Different cooks respond to this knowledge in different ways. Some fight the machine, seeking out an artisanal cuisine, which asserts with every rugged bite that the meal has been made by hand. Many Italian families even now happily spend hours rolling, cutting, and pinching tortelloni by hand, because the factory-made versions of filled pasta—unlike the best machine-made dried pasta, which is unimprovable—cannot compete with homemade. Yet they do not go so far as to take out a pestle and grind their own flour for the pasta. The cult of handmade food only goes so far, because we all have better things to do than spend hours grinding.

The Slow Food Movement started in Italy in 1989 to "counter the rise of fast food and fast life." The slowness of Slow Food refers primarily to methods of agriculture and ways of eating: its philosophy defends biodiversity against intensive farming, and slow and sensuous meals as opposed to quickly grabbed bites. The movement also favors food that is slow to produce. Slow Food goes along with a cult of the handmade and the homemade as against the machine-made, with a rediscovery of the therapeutic pleasures of kneading sourdough or making home-cured salami from scratch, recreating as leisure pursuits what were once backbreaking kitchen work.

But slow and difficult is not the only way to make delicious food. Other, more pragmatic cooks embrace the machine. The great chef Raymond Blanc sometimes demonstrates how to make sweet pastry in a food processor, tipping in butter, flour, and sugar and pulsing for no more than half a minute with egg yolk and water, forming it with a dexterous movement into a ball of buttery dough. "You can make it by hand if you like," I once heard him comment, reasonably enough. "But it will take much longer and it won't be any better."

~ Nutmeg Grater ~

FORM FOLLOWS FUNCTION. JUST LOOK AT THE difference between a British nutmeg grater and a Japanese ginger grater. One is metal, with punched-out perforations, to rasp the nutmeg to a fine powder. The other is a ceramic dish, with spikes not holes. The spikes trap the fibrous parts of the ginger, while the flavorsome juice and pulp run out to the sides.

In their different ways, both are very gratifying tools. They owe their existence to the passion for warming spice and the circuitous twists of trade, agriculture, and taste that embed one spice or another in a nation's cuisine.

Nutmeg, harvested on the Spice Islands of Indonesia, was the most feverishly desired luxury in seventeenth-century Europe. Nutmeg pomanders were used against the plague, and the soothing, mildly hallucinogenic spice was grated into dishes both savory and sweet. It is no longer such a central British taste, apart from its use in eggnog, Christmas puddings, bread sauce, and custard tarts. People now do not carry their own personal supply with them in tiny boxes. Yet it is the one spice that we still insist on grating fresh, sometimes keeping the little brown ovoids snapped shut inside half-cylindrical graters that look just like nutmeg graters have always looked.

Japanese cooking is not spice heavy. But ginger is crucial, whether pickled pink ginger to go with sushi or the fresh rhizome grated into sauces with soy and sake. Ginger is just one of the fibrous plants a Japanese cook needs to deal with—others include wasabi and daikon radish. Early Japanese graters were often made from shark skin, to catch the stringy bristles. Now, ceramic bumps do the job instead, like a vicious form of Braille.

You can't grate ginger on a nutmeg grater—the wet root quickly clogs up the holes in the metal. Nor can you grate nutmeg on a ginger grater—the hard spice slips off the spikes, hurting your hands. If you need a tool to grate both spices (and zest lemon, and grate Parmesan), then forget tradition and buy a Microplane.

❧ 6 ❧

EAT

Have your table linen sweet and clean,
your knives bright, spoons well washed.
JOHN RUSSELL, *The Book of Nurture* (c. 1460)

Fingers were made before knives and forks.
Old adage

S POONS—ALONG WITH THEIR COMPANIONS AND RIVALS, chopsticks and forks—are definitely a form of technology. Their functions include serving, measuring, and conveying food from plate to mouth; not to mention culinary spoons for stirring and scraping, skimming, lifting, and ladling. Every human society has spoons of one kind or another. In and of themselves, these utensils are mild-mannered—certainly in comparison with the knife. Spoons are what we give babies—whether ceremonial silver christening spoons or shallow plastic weaning spoons containing the first gummy mouthfuls of baby rice. Gripping a spoon in the fist is one of the earliest milestones in our development. Spoons are benign and domestic. Yet their construction and use has often reflected deep passions and fiercely held prejudices.

I n 1660, the luxuriantly bewigged Charles II became king of England, Scotland, and Ireland, a restoration of the monarchy after the country's brief experiment with republican government in the Commonwealth of Oliver Cromwell and his son Richard. Eleven years earlier, in 1649, the king's father, Charles I, was executed, the culmination of the English Civil War. Now monarchy was back with a vengeance. Charles II's Restoration was accompanied by sweeping cultural changes, aimed at effacing all memory of the Puritan Roundheads. Theaters reopened. Handel composed his majestic *Water Music*. And, almost overnight, silver spoons took on an entirely new shape, the trifid (also known as trefids, trefoils, split-ends, and pieds-de-biche).

Because the Commonwealth lasted such a short time, Cromwellian spoons are rare. But those that have survived are, as you'd expect, plain and unadorned. The shape of these spoons—which began to appear in England from the 1630s on—is known as "Puritan." They have a simple, shallow egg-shaped bowl that gives way to a plain, flat stem. The Puritan spoon marked a departure from previous English silver spoons, which had bowls that were fig shaped (the technical term is ficulate), with chunky hexagonal stems. These earlier spoons had a bowl like a teardrop, widening toward the end that you put in your mouth, whereas the Puritan bowl narrowed slightly at the end, like most of our spoons now.

The biggest change with the Puritan spoon was its handle, which was entirely unadorned. It had no decorative "knop" on the end. Over the previous few centuries, silversmiths lavished great artistry on a part of the spoon we would now consider almost irrelevant, adding little sculptures called knops on the end point of the handle. Pre-1649 knop "finials" included diamonds and acorns, owls and bunches of grapes, naked women and sitting lions. Some knops were flat-ended abstract shapes, such as a stamp or a seal. Others depicted Christ and his apostles in ornate finials.

None of these decorative spoons found favor during the Commonwealth, when excessive decoration of any kind, particularly religious, was disapproved of. The Roundheads lopped the heads off spoons just as they lopped off the king's head. The new republican eating utensils were entirely devoid of pattern, just plain, dense lumps of silver. It has been suggested that one reason Puritan spoons were made so heavy was that citizens used them to hoard silver against the frequent proclamations that came through to give up your personal silver to pay for the defense of the town. If your silver was tied up in cutlery, you could claim it was essential and prevent its being confiscated.

In any case, it wouldn't be long before the Puritan spoon was itself swept away by the spoon of the Restoration, the trifid, which traveled with the newly crowned Charles II from his court of exile on the Continent. It is the earliest spoon in its modern form; most spoons today, however cheaply made, still owe something to the trifid. No British person had ever eaten from such a spoon before in Britain—the first trifids are hallmarked 1660. Yet by 1680, they had spread through the entirety of Charles's kingdom and remained the dominant spoon type for forty years, killing off both the Puritan spoon and the fig-shaped spoons that went before. The base metal spoons of the masses made from pewter and latten also changed shape from Puritan to trifid. The change was not gradual, but sudden. Politically, no one wanted to be seen eating dinner with a Roundhead spoon.

The bowl of the trifid was a deep oval rather than a shallow fig. Like the Puritan, the trifid had a flat handle, but it now swelled toward the end, with a distinctive cleft shape (hence the name, which means "three-cleft"). The design is French; the trefoil is an echo of the fleur-de-lis, the stylized lily associated with French kingship. On the reverse side, the hammered stem continued up onto the back of the bowl, finishing in a dart-shaped groove sometimes called a "rat tail." Over the decades, these new spoons also seem to have gone along with changes in the way they were held. Certain shapes invite you to hold them in certain ways. Because of the knobbly part at the end, medieval spoons are easiest to hold with the stem under the thumb at a right angle. The trifid, by contrast, could be held in the polite English way, with the handle resting in the palm of the hand, parallel with the thumb. With a regal trifid in your hand, poised to plunge it into an apple pie, you might forget that a reigning monarch had ever been executed or that England had ever done without its king. This was kitchenware as political propaganda.

Spoons hold up a mirror to the surrounding culture precisely because they are so universal. There are fork cultures and there are chopstick cultures; but all the peoples of the world use spoons. The particular form they take is therefore very revealing: a pretty Chinese porcelain blue-and-white spoon for wonton soup is part of an entirely different culture of eating than a Russian spoon filled with sticky preserves or the ladle-like wooden spoons used in poor European households to eat soup from a communal pot, passed from mouth to mouth. Functionally, a spoon is an object that aids with ferrying food into the mouth. In the 1960s, Jane Goodall saw chimpanzees fashioning sort-of spoons from blades of grass, to make it easier for them to slurp up termites. In the distant past, humans lashed shells onto sticks and used them to consume foods too liquid to be eaten with fingers. The Roman word for spoon reflects this: *cochleare*, which comes from the word for shell. Romans used these little spoons for eating eggs or scooping out shellfish. For pottage-type dishes they had a larger spoon, the pear-shaped *ligula*.

At different periods, people favored various spoons, depending on what they most liked to eat. Mother-of-pearl egg spoons reflect the Edwardian fondness for a soft-boiled egg (mother-of-pearl or bone was used because egg yolk stains silver). Hanoverian mustard spoons hint at what a vital condiment this fiery fluid was in the English diet. The Georgians of the eighteenth century loved roasted bone marrow and devised a series of specialized silver spoons and scoops for eating it. Some of them were double-ended, with one end for small bones and another for large. The idea was to hold your piece of roasted bone in an elegant white napkin and use the implements to tease out the soft, fatty nuggets of marrow. Marrow spoons were akin to the complicated series of spoons, needles, and picks that accompany a French *plateau de fruits de mer,* a sumptuous seafood platter.

The marrow spoon is obsolete (though the fashion for roasted bone marrow and parsley salad recently started by the London chef Fergus Henderson may yet bring it back). Other spoons, however, have succeeded in making the leap from specialist tool to universal implement, none more so than the teaspoon. The teaspoon first came into existence when the English started adding milk to their tea in the second half of the seventeenth century. It was needed to amalgamate the milk, sugar, and tea in the cup. It was a wealthy person's utensil, separate from the main dinnerware. On the face of it, it seems odd that the teaspoon should have made the leap from the rarefied atmosphere of an English tea table to cutlery drawers around the world. The utensils of the Japanese tea ceremony—the bamboo tea scoop and whisk—have not traveled in this way. Nor have other accoutrements of an English tea—such as sugar tongs and tea strainers, which have remained the preserve of those who still enjoy the ritual of stopping for a full bone-china afternoon tea, complete with scones and cream: a dwindling band. You only rarely encounter anyone polite enough to use tea tongs now, not least because sugar cubes themselves have passed out of vogue. Yet teaspoons are still everywhere.

The teaspoon did not immediately travel the world. In 1741, the inventory of the French duc d'Orléans included forty-four silver-gilt coffee spoons but not a single teaspoon. The French are still likely to use the smaller coffee spoon as a unit of measurement over the teaspoon (abbreviating it to cc, short for *cuiller à café*). But elsewhere the teaspoon reigns supreme, even when tea itself is not drunk. From the nineteenth century onward, the teaspoon became a basic element of flatware in the United States, despite the fact that coffee was more usually drunk; and thence, its influence spread. But why? How did the teaspoon make the leap into the mainstream culture while other specialist spoons did not, such as the Victorian berry spoon with its lacy trim or the small silver salt shovels that were made in profusion in the eighteenth century, some like mini-soupspoons, others like tiny ice-cream servers?

I suspect the reasons for the global success of the teaspoon are twofold. First, in its primary function, it is not really a spoon for tea, but a spoon for sugar, a substance that is just as popular among coffee drinkers as tea drinkers. Second, the teaspoon answered a genuine need for a handy little utensil, smaller than either the eighteenth-century tablespoon or dessertspoon, but neither so tiny as the French coffee spoon nor as fussy as a Georgian salt shovel. An American teaspoon was larger than an English one, but in either case, the dimensions were a pretty good fit for the human mouth. The teaspoon's uses are myriad, as demonstrated by the tendency they have to disappear from the cutlery drawer (only kitchen scissors are more elusive). They are constantly called into service for measuring small quantities of baking powder and spices. Most cooks use them as tasting spoons, too, dipping them into a sauce to gauge the seasoning, or just to sneak a pleasurable foretaste of dinner. And then there are all the things you can eat handily with teaspoons, from little cup custards to avocados. I am rather biased on this account because as an eccentric and somewhat troubled teenager, for several years I ate all my food—anything that didn't need cutting, at any rate—with a teaspoon. Clearly, I had some unresolved "issues." I remember how safe it made me feel, spooning up tiny morsels like an infant.

So, in extremis, one spoon can definitely be used for all meals, which is not to say that it will work equally well for all foods. Because the end result is always the same—getting food in your mouth—it is seldom recognized that an eating spoon can operate in at least two different ways. The bowl of the spoon can be a kind of cup, from whose edge liquids are drunk. Or it can be a shovel, designed for ferrying more solid mixtures. A very clear example of the spoon-as-shovel is the *kafgeer,* a large flat spoon used in Afghanistan for serving rice, rather like a spade. Throughout the Middle East, there are special shovels and spatulas for serving rice, and when you use them, you notice that they really do a much better job of picking up every grain than our rounded, oval spoons.

Similarly, when you look at early European spoons, you can detect radical differences in shape, reflecting different usage. From a nunnery on the remote Scottish island of Iona, medieval silver spoons have survived that have a distinctive leaf-shaped bowl: definitely a shovel, though a much smaller one than the Middle Eastern rice servers. These spoons would have been ideal for scooping up thick porridge, but not much good for liquid soup. For this, medieval spoon makers made large round spoons, whose bowls were too big to fit in the mouth, but just right for sipping.

Now, mostly, we don't think too hard about how spoons work. The reason for this is partly that the modern spoon with its ovoid bowl marks a compromise between the cup and the shovel. Pick a dessertspoon from your cutlery drawer. Could you use it to scoop up mouthfuls of pilaf, say? And could you use it to drink thin broth? The answer to both these questions should be yes. Your dessertspoon is probably not perfect for either task: too shallow for soup, too deep and rounded for rice. But it will do. For John Emery, this compromise was not good enough. Emery was a spoon fanatic, a historian of cutlery who experimented in the 1970s with making replicas of historic spoons and testing what could and couldn't be eaten with them. From the perspective of function, he lamented the trifid and all its successors. In Emery's view, the compromise between cup and shovel was "seldom really satisfactory." And it was made still worse by the annoying habit food had of oscillating between the states of solid and liquid. Sometimes soup was thick and lumpy like porridge, and sometimes porridge was thin like soup. Etiquette told Emery to use one spoon; function, another.

For Emery, as for all spoon connoisseurs, the answer is evermore-specialized spoons. If you were of this mindset, you'd have been in heaven in Victorian times, when there were aspic spoons and tomato spoons, sauce spoons and olive spoons, fluted gravy ladles, bon-bon spades, tea scoops, citrus spoons, and Stilton scoops, among others. The proliferation of flatware was fueled by the move from service à la française (all the dishes were set on the table at once for diners to help themselves) to service à la russe (dinner was served in a succession of courses, each requiring its own utensils). Late nineteenth-century America saw an even greater range of new refined spoons: not just rounded soupspoons (first introduced in the 1860s) but distinct spoons for cream soups and bouillons (the latter were smaller). And the serving spoons! Implements included special servers for fried oysters, chipped beef, macaroni, and potato chips. Tiffany's marketed a solid silver "Saratoga Chip Server," named after Saratoga Springs, New York, where the potato chip was first served; the implement had a stubby handle and a ballooning openwork bowl, to preserve genteel hands from the horror of handling fried potatoes. It is not clear, however, that this proliferation in eating and serving tools was a sign of progress.

H aving more gadgetry—more kitchen belongings—at our disposal does not necessarily make life easier. The problem with assembling more and more shiny tools that deal with the messy business of cooking and eating is that they tend to arrive with social mores that deem it necessary to use the tools, even when it flies in the face of common sense to do so. Food writer Darra Goldstein speaks of "fork anxiety," a nervous condition brought on by the smorgasbord of silverware laid out on the table of grand dinners: "There probably never was a time when all of these forks were in use, but you can see how the very existence of a tomato fork could generate anxiety," noted Goldstein in 2006. Etiquette books of the early twentieth century go to elaborate lengths to describe how to use silver knives and forks to deal with foods that would be far better picked up and eaten with fingers: ripe peaches, corn on the cob, anything with bones.

Most of the polite rules surrounding cutlery reflect a terror about handling food—an anxiety about its stickiness and noise. Again and again, we are told that soup must be sipped silently—in contrast to Japan, where the etiquette for noodle soup eating decrees that it be noisily sucked and slurped, to demonstrate true enjoyment. In the West, protocol said it must be drunk from the edge of the spoon—it was thought ill-mannered to insert too much of a spoon into your mouth—though a special dispensation was created for men with full moustaches, who were allowed to drink soup from the end of the spoon. In 1836, it was thought that to pick up sugar using fingers rather than sugar tongs was such a terrible faux pas, it might lead to a gentleman losing his good reputation. On the other hand, there was also an anxiety about seeming too refined or minding too much about the finer points of table manners. To go on too much about the right fork was a sign of insecurity or even fraudulence. Real aristocrats knew the "refined coarseness" of when to employ fingers instead of a fork: fingers were right for radishes, crackers, celery, unhulled strawberries, and olives. A fictitious story circulated of an adventurer who tried to pass himself off as a nobleman. Cardinal Richelieu uncovered this rascal when he attempted to eat olives with a fork, something no true gentleman would do.

The use of knives, forks, and spoons is part of a wider culture of manners and a larger civilization of conformity. Although it might not have mattered too much if you used the wrong fork, it was essential to demonstrate that you understood the rules of the game. The key was to act as if you belonged. This was the hardest thing of all, especially because fashions in the use of tableware changed rapidly, and a custom that was de rigeuer one decade could become ridiculous the next. In the early nineteenth century, there was even a brief vogue among "fashionables" for eating soup with a fork. It was soon condemned as "foolish," and the spoon was restored.

But for almost everything else, the politest way to eat was still with a fork. Among the English upper classes of the mid-twentieth century, the "fork luncheon" and the "fork dinner" were buffet meals at which the knife was dispensed with altogether. The fork was polite because it was less overtly violent than a knife, less babyish and messy than a spoon. Forks were advised for everything from fish to mashed potatoes, from green beans to cream cake. Special forks were devised for ice cream and salads, for sardines and terrapins. The basic rule of Western table manners in the nineteenth and twentieth centuries was: if in doubt, use a fork. "Spoons are sometimes used with firm puddings," noted a cookbook of 1887, "but forks are the better style."

Yet we have short memories when it comes to manners. It was not so long ago that eating anything from a fork had seemed absurd. As a kitchen tool, the fork is ancient. Roasting forks—long spikes for prodding and lifting meat as it cooks—have been around since Homeric times. Carving forks, to hold meat down as it is cut, are medieval. Yet forks for eating as opposed to forks for food preparation only started to seem a good idea in the modern era. The table fork is far less time-honored than such objects as the colander, the waffle iron, the bain-marie. In the great scheme of things, eating with prongs is a novelty.

In the parts of the world where forks are not used, they seem profoundly alien instruments—little metal spears that, unlike chopsticks or fingers, clash with food as it enters the mouth. Yet in the West, we use them so universally, we think nothing of them.

In the contemporary Western world, unless we are eating sandwiches or soup, almost every meal entails a fork. We use the fork to spear we eat now vegetables and to steady meat as we cut it; to pick food up or to chase it around the plate; to twirl spaghetti; to flake fish; to build up fragments of different foods into a single choice mouthful; or to hide pieces of unwanted cabbage from our parents' beady eyes. Children play with forks, using the sharp tines to reduce green beans to a mush, or to turn potatoes pink with ketchup. In a formal mood, we may even use a fork to eat a slice of cake, crumb by crumb. At fancy dinners or weddings, we still worry about which ornate fork to use for which course, but forks are also found at the most casual of meals, for the kind of basic snacks for which knives would be out of place. Office workers sit in the park eating pasta salad with a disposable fork, with half an eye on the crossword. Even kebab-eaters, reeling from the pub, will grasp a plastic fork to spare their fingers from the grease.

We take forks for granted. But the table fork is a relatively recent invention, and it attracted scorn and laughter when it first appeared. Its image was not helped by its associations with the Devil and his pitchfork. The first true fork on the historical record was a two-pronged gold one used by a Byzantine princess who married the doge of Venice in the eleventh century. St. Peter Damian damned her for "excessive delicacy" in preferring such a rarefied implement to her God-given hands. The story of this absurd princess and her ridiculous fork was still being told in church circles two hundred years later. Sometimes the tale was embellished. The princess died of plague: a punishment, it was said, for eating with a fork.

Six centuries later, forks were still a joke. In 1605, the French satirist Thomas Artus published a strange book called *The Island of Hermaphrodites*. Written during the reign of Henri IV, it made fun of the effeminate ways of the previous monarch, Henri III, and his court of mollycoddled hangers-on. In the sixteenth century, "hermaphrodite" was a pejorative term, which might be applied to anyone you didn't much like. In mocking these courtiers, one of the worst things Artus could think of was that they "never touch meat with their hands but with forks," whose prongs were so wide apart that the hermaphrodites spilled more broad beans and peas than they picked up, scattering them everywhere. "They would rather touch their mouths with their little forked instruments than with their fingers." The implication is that using forks was—like being a hermaphrodite—a kind of sexual abnormality. To Artus, the fork was not just useless—it was obscene.

It was not that spiky fork-like instruments were unheard of before then, but their use was limited to certain foods. In ancient Rome, there were one-pronged spears and spikes for getting at hard-to-reach shellfish, for lifting food from the fire or impaling it. Medieval and Tudor diners also had tiny "sucket" forks, double-ended implements with a spoon at one end and a two-pronged fork at the other. As sugary sweetmeats or "suckets" became more common among the rich, so the need for these forks increased. In 1463, a gentleman of Bury St. Edmunds bequeathed to a friend "my silvir forke for grene gyngour" (candied ginger). The fork end was used to lift sticky sweetmeats out of pottery jars; the spoon end was used to scoop up the luscious syrup. When bits of sweetmeat lodged in the teeth, the sucket fork doubled as a nifty toothpick. But this was not at all the same as a fork in the modern sense—an individual instrument to enable people to eat an entire meal without handling the food.

Forks in our sense were considered odd until the seventeenth century, except among Italians. Why did Italy adopt the fork before any other country in Europe? One word: pasta. By the Middle Ages, the trade in macaroni and vermicelli was already well established. Initially, the longer noodle-type pastas were eaten with a long wooden spike called a *punteruolo*. But if one spike was good for twirling slippery threads of pasta, two were better, and three ideal. Pasta and the fork seem made for one another. It is a joy to watch a table of Italians eating long ribbons of tagliatelle or fettuccine, expertly winding up forkfuls of pasta, like slippery balls of yarn. Having discovered how useful forks were for eating noodles, Italians started to use them for the rest of the meal, too.

When Thomas Coryate, an Elizabethan traveler, journeyed around Italy sometime before 1608, he noticed a custom "that is not used in any other country," namely, a "little fork" for holding meat as it was cut. The typical Italian, noted Coryate, "cannot endure to have his dish touched with the fingers, seeing all men's fingers are not clean alike." Although it seemed strange to him at first, Coryate acquired the habit himself and continued to use a fork for meat on his return to England. His friends—who included the playwright Ben Jonson and the poet John Donne—in their "merry humour" teased him for this curious Italian habit, calling him "furcifer" (which meant "fork-holder," but also "rascal"). Queen Elizabeth I owned forks for sweetmeats but chose to use her fingers instead, finding the spearing motion to be crude.

In the 1970s, real men were said not to eat quiche. In the 1610s, they didn't use forks. "We need no forks to make hay with our mouths, to throw our meat into them," noted the poet Nicholas Breton in 1618. On the cusp of the twentieth century, as late as 1897, British sailors were still demonstrating their manliness by eating without forks. This was a throwback, for by then forks were nearly universal.

By 1700, a hundred years after Coryate's trip to Italy, forks were accepted throughout Europe. Even Puritans used them. In 1659, Richard Cromwell, the lord protector, paid 2 pounds and 8 shillings for six meat forks. With the Restoration, forks were firmly established on the table, alongside the new trifid spoons. Not wanting to dirty your fingers with food, or to dirty food with your fingers, had become the polite thing to do. The fork had triumphed, though knives and spoons continued to outsell forks until the early nineteenth century.

The triumph of the knife and fork went along with the gradual transition to using china dinner plates, which were generally flatter and shallower than older dishes and trenchers. When bowls were used for all meals, the ideal implement was a spoon with an angled handle for digging deep, like a ladle (the fig-shaped spoons of the Middle Ages usually had stems pointing upward). A knife or fork with horizontal handles does not sit naturally in the curved structure of a trencher or a pottage bowl. They need a flat surface. Try to eat something in a deep cereal bowl using a knife and fork, and you will see what I mean; your elbows hunch up and your ability to use the cutlery is severely restricted. Flatness is also necessary for the elaborate semaphore of knife-and-fork table manners that reached their apogee in Victorian times. The plate becomes like a dial, on which you communicate your intentions.

It is sometimes said that the earliest forks were all two-pronged. This is not so. Some very early forks have survived with four prongs (or "tines"), others with three, and a greater number with two. The number of prongs was an indication not of date but of function. Two prongs were best suited to impaling and stabilizing food—mostly meat—while it was cut (like the carving forks still sold as a set with carving knives). Three prongs or more were better if the fork was to be used as a quasi-spoon, for conveying food from plate to mouth. There were even experiments to push it to the limit with five-pronged forks (like the five-bladed razors that took over from the old two-bladed and three-bladed ones, claiming hyperbolically to be the most "technologically advanced" way for men to shave), but this was found to be too much metal for the human mouth to hold.

In the nineteenth century, two distinct methods emerged for handling a knife and fork. The first was christened by the great etiquette guru Emily Post as "zigzag" eating. The idea was to hold your knife in the right hand and fork in the left as you cut up everything on the plate into tiny morsels. You then put the knife down, seized the fork in the right hand, and used it to "zigzag" around the plate, scooping up all the morsels. At first, this method was common throughout Europe, but it later came to be seen as an Americanism, because the English devised a still more refined approach. In English table manners, the knife is never laid down until the course is finished. Knife and fork push against one another rhythmically on the plate, like oars on a boat. The fork impales; the knife cuts. The knife pushes; the fork carries. It is a stately dance, whose aim is to slow down the unseemly business of mastication. Both the Americans and the British secretly find each other's way of using a fork to be very vulgar: the British think they are polite because they never put down their knives; Americans think they are polite because they do. We are two nations separated by common tableware, as well as by a common language.

In the four hundred years since Thomas Coryate marveled at Italian meat-forks, our food has changed immeasurably, yet our dependence on the fork largely has not; we use them more now than ever. Like the colander, in use since ancient times, it is an example of a kitchen technology that has stuck. Although we may abandon it to munch a hamburger or to attempt to use chopsticks in a Chinese restaurant, the fork is entirely bound up with our experience of eating. We are so used to the sensation of metal (or plastic) tines entering our mouths along with food, we no longer think anything of it. But our use of forks is not inconsequential—it affects our entire culinary universe. As Karl Marx observed in the *Grundrisse*, "The hunger gratified by cooked meat eaten with a knife and fork is a different hunger from that which bolts down raw meat with the aid of hand, nail and tooth." Forks change not just the *how* of eating but the *what*.

Which is not to say that forks are always superior to other methods of eating. As with every new kitchen technology from fire to refrigeration, from eggbeaters to microwaves, forks have drawbacks as well as benefits. The Renaissance opponents of the fork were right, in many ways. Knives and forks are handy enough for cutting a slice of roast beef, but are more hindrance than help for eating peas or rice, which are better served by the humble spoon. Eating with a knife and fork carries with it a complacency that is not always justified. It is a very fussy way of eating food. We often overattribute efficiency to the technologies we are accustomed to. Because we use knives and forks every day, we do not notice how they hamper us. Our table manners require us to use two hands to perform with less dexterity what chopsticks can do with only one.

Monkies with knitting needles would not have looked more ludi-crous than some of us did," commented one of those present on the first recorded occasion of Americans eating Chinese food in China, in 1819. Chinese hosts in Guanzhou were entertaining a party of American traders. A procession of servants brought in a series of "stewed messes" and dishes of bird's-nest soup, plus plenty of boiled rice. "But alas!" recalled one young trader from Salem, Massachusetts, "no plates and knives and forks." The Americans struggled to ingest any of the feast with the sticks provided until at last their hosts took pity and ordered knives, forks, and spoons to be brought.

There is sometimes a similar moment when Westerners eat out at a Chinese restaurant. Halfway through dinner, you notice that some-one is silently blushing because he has no idea how to use chopsticks and is struggling to get anything in his mouth. It takes tact on the part of the restaurateur to rustle up a spoon and fork without making the customer feel stupid. A Chinese woman who settled near Harvard in the 1950s noted that when entertaining Americans, it was important to have forks ready for emergencies, but also important not to press them on guests who insisted on practicing ineptly with chopsticks. The Western knife-and-fork eater faced with chopsticks for the first time is reduced to the level of a clumsy child. The ability to use chopsticks is like literacy, a serious skill, not easy to master, but essential to being a fully functioning member of society in China, Japan, or Korea. For the first few years of a child's life in China, it is fine to use a spoon. After that, a child may have chopsticks joined together with napkins and rubber bands, to form a kind of makeshift tongs. But on reaching sec-ondary school age, the window of forgiveness has passed. You are now expected to know how to wield your chopsticks dexterously. To fail to do so would be taken as a sign of bad parenting.

The earliest pair of surviving chopsticks are bronze, from the Ruins of Yin and dated around 1200 BC, so we know that they have been in use for at least 3,000 years. But it was only from around the Han dynasty (206 BC–AD 220) that they became the universal method of eating all over China. The rich had chopsticks made of bronze, ivory, jade, or finely painted lacquer; the poor had simple wooden and bamboo ones. At the imperial table, silver chopsticks were used, not just for their luxury but to aid in the detection of poison: the idea was that the silver would turn black if it came into contact with arsenic. The downside was that silver is heavy, conductive of heat (becoming too hot when in contact with hot food and too cold in cold food) and—this is a pretty basic flaw—bad at picking up food (the silver does not provide enough friction, making them slippery). Eventually, therefore, silver chopsticks were abandoned, despite their beauty and poison-detecting potential, because they violated one of the most basic aspects of Far Eastern table manners: the duty to demonstrate enjoyment of the deliciousness of what was on the table. It was easier to show pleasure with porcelain chopsticks.

As seen in Chapter 1, the use of chopsticks went along with an entirely different approach to cuisine than in Western cooking. Because chopsticks only lift food rather than chopping it, all the knife work could be hidden away in the kitchen. "Everything is served cut up," noted Fletcher Webster in 1845, another American traveler to China. The chopping skills of the cook thus saved those at the table from all the worries a Western diner faces about how to subdivide the food on his or her plate without looking uncouth. How to eat corn on the cob politely was not a dilemma faced by any Chinese eater, not just because corn was not grown in China, but because for the cook to lump such a large object on the plate would itself have been unimaginably rude.

The system of eating with chopsticks eliminates the main Western taboos at table, which chiefly have to do with managing the violence of the knife. The French theorist Roland Barthes, who saw symbols everywhere but especially at the table, argued that chopsticks were the polar opposite of the knife. Holding a knife makes us treat our food as prey, thought Barthes: we sit at dinner ready "to cut, to pierce, to mutilate." Chopsticks, by contrast, had something "maternal" about them. In expert hands, these sticks handle food gently, like a child:

> The instrument never pierces, cuts or slits, never wounds but only selects, turns, shifts. For the chopsticks . . . never violate the food-stuff: either they gradually unravel it (in the case of vegetables) or else prod it into separate pieces (in the case of fish, eels), thereby rediscovering the natural fissures of the substance (in this, much closer to the primitive finger than to the knife).

Despite their basic gentleness, however, it is still possible to cause offense when eating with chopsticks. Superficially, Chinese table manners are more relaxed than traditional European and American ones: the table setting consists of nothing but a single pair of chopsticks and a three-piece porcelain set consisting of a spoon, a bowl, and a small plate. When Florence Codrington, a British woman in China in the early twentieth century, invited a Chinese "old lady friend" to her house to eat dinner English style, she "ran round and round the table in wild excitement, touching everything in turn and then held her sides laughing. 'Ai-a! it is laughable, it is surprising!' she gasped, 'all these tools to eat a meal with!!'" Unlike the traditional Western procession of individual courses, Chinese dishes are set on the table to share communally, with everyone picking at the same time. It is not rude to reach across others for a far-off dish. Chinese food writer Yan-Kit So observed that the "likelihood of chopsticks clashing is minimal."

On the other hand, because Chinese cuisine is part of a culture of frugality, there are strict rules about eating food in such a way that neither waste nor the appearance of waste is allowed, particularly when it comes to rice. The way in which everyone shares dishes may seem random, but a mark of good manners is that no one present should be able to tell what your favorite dish is; in other words, you should not dig your chopsticks greedily into the same dish too often. As for rice eating, the bowl should be raised to the lips with the left hand, while the chopsticks shovel the rice in. Every last grain of rice must be eaten. British children who leave food on their plates are warned to think of the starving in Africa. Chinese children—who eat several small helpings from a bowl instead of a single heaping plateful—are given a different, more persuasive, admonition against wastefulness: to think of the sweat on the brow of the farmer who grew their rice.

The Japanese came to chopstick culture later than the Chinese (from whom they borrowed the idea), but you would not know it now, from the way that chopsticks shape the entire culinary universe of the country. It was only around the eighth century AD that chopsticks supplanted hands among the common people, but having done so, they rapidly became essential to the Japanese way of eating. Japanese chopsticks tend to be shorter than Chinese ones (around 22 cm as against 26 cm), and they have pointed ends rather than flat ones, allowing for the most minute specks of food to be picked up. If a food can neither be eaten with chopsticks nor drunk from a bowl, it used to be said, then it is not Japanese. As Japanese food has become globalized in recent decades, this rule no longer holds, entirely. Among the young of Tokyo and Osaka, two of the most popular dishes are breaded pork *katsu* cutlet—which is usually sliced on the diagonal in slices that require further cutting with a knife—and "curry," a strange all-purpose spicy sauce, gloppy and redolent of canteen food, which many in Japan adore. This curry cannot be eaten with chopsticks and is too thick to drink from a bowl: it calls for a spoon. Another popular Japanese food is the white bread "sando," imitations of the British sandwich, made from sliced white bread stuffed with mayo-heavy fillings. It is held, as every true sandwich must be, in the hand.

Nevertheless, what is eaten and how it is eaten in Japan are still largely shaped by chopsticks, and there are a series of very specific forms of behavior that must be avoided when holding them.

In addition to the obvious taboos against using chopsticks in such a way as to suggest violence—pointing them in someone's face, sticking them upright in a plate of food—there are more subtle transgressions. These include:

Namida-bashi (crying chopsticks): letting a liquid drip like tears from the end

Mayoi-bashi (hesitating chopsticks): allowing your chopsticks to hover over various dishes of food without choosing between them

Yoko-bashi (scooping chopsticks): using chopsticks like a spoon

Sashi-bashi (piercing chopsticks): using chopsticks like a knife

Neburi-bashi (licked chopsticks): licking off fragments from the end of the chopsticks

There are also taboos about the sharing of chopsticks. The Shinto religion has a horror of impurity or defilement of any kind. There is a belief that something that has been in someone else's mouth picks up not merely germs, which would be killed by washing, but aspects of their personality, which would not. To use a stranger's chopsticks is therefore spiritually disgusting, even when they have been washed. Professor Naomichi Ishige is an anthropologist of Japanese food who has published over eighty books. He once conducted an experiment on some of his Japanese seminar students, asking them: "Suppose you lend an article that you use to someone else, who uses it, and then thoroughly cleans it before returning it to you. Which article would you have the strongest sense of psychological resistance against reusing afterward?" The two objects that most often cited were a pair of undergarments "for the lower part of the body" and a pair of chopsticks.

This goes some way toward explaining the phenomenon of *waribashi*, disposable chopsticks made from a small piece of cheap wood almost split in half, ready for the customer to pull apart and use. It is sometimes assumed that these *waribashi* are a modern Western influx, akin to polystyrene cups. But this is not so: they have been used ever since the beginnings of the Japanese restaurant industry in the eighteenth century, because giving a fresh pair of chopsticks to all customers was the only way a restaurateur could assure his clientele that what they put in their mouth was not defiled. They are a good example of how what we are prepared to accept in the way of the technology of eating is often determined more by cultural forces than function. Richard Hosking, a British expert on Japanese food, argues that "from the point of view of a foreigner not entirely at ease with chopsticks, waribashi are wretched," because their shortness makes them tricky for large-handed people. They also have an annoying tendency to splinter apart the wrong way, forcing you to undergo the embarrassment of asking for another pair. Worse than that, *waribashi* are an ecological disaster. Japan now uses and throws away around 23 billion pairs per year.

The appetite for disposable chopsticks, moreover, has spread to China, which now manufactures 63 billion pairs annually. By 2011, the Chinese demand for disposable wooden chopsticks was so great that it could no longer supply enough of the right kind of wood for its 1.3 billion citizens. An American manufacturing plant in Georgia has started to plug the gap. The state of Georgia is rich in poplar and sweet gum trees, whose wood is pliable and light enough to need no bleaching before it is made into chopsticks. The company, Georgia Chopsticks, now exports billions of disposable chopsticks to supermarket chains in China, Japan, and Korea, all with a label stating "Made in U.S.A."

Those first American traders who arrived in China in the nineteenth century and struggled with chopsticks like "monkies with knitting needles" probably never thought the day would come when the United States would be supplying chopsticks to China. In the end, however, the two cultures—the knife-and-spoon culture and the chopstick culture—have more in common than appears at first.

When dining with one another, each of them may have secretly thought: *You barbarians!* But both cultures are united in their disdain for a third group, those who manage the business of eating without any tools at all.

P rejudices are by definition not reasonable, so perhaps we should not be surprised that most of the prejudices against eating with fingers turn out to have little basis in fact, when examined closely. First, there is the notion that touching food is a sign of slovenliness; second, that eating with fingers demonstrates a lack of manners; and third, that having no eating utensils limits what can be eaten. The answer to these concerns are: (1) No, (2) No, and (3) Only sometimes.

Lack of cutlery does not signal lack of manners. Among people who consistently eat with fingers, performing elaborate ablutions becomes part of the rhythm of the meal. Even King Henry VIII, whose eating with fingers has become a byword for gross table manners, was actually far more attentive to both hygiene and etiquette than most sandwich eaters today. The king's carver cleared away any crumbs using a voiding knife. Ushers provided him with napkins and swept specks of food off his clothing. At the end of the meal, a nobleman knelt before him with a basin, so that he could wash all traces of food off his hands. We may joke about Henry's revolting manners, but how many of us stay half as clean during mealtimes?

A cultural preference for eating with fingers tends to make diners very sensitive to cleanliness. Ancient Romans washed themselves from top to toe before dinner. Arabs in the desert rub their hands with sand. Many Arabs today use a fork and spoon, but before a traditional Middle Eastern meal, writes Claudia Roden, guests are entertained on sofas, where their hands are cleaned:

> A maid comes round with a large copper basin and flask, pouring out water (sometimes lightly perfumed with rose or orange blossom) for the guests to wash their hands. A towel is passed round at the same time.

In the ninth century, among Arabs, if a single guest so much as scratched his head after washing, everyone at the table would have to wait for him to wash all over again before they started to eat. The little finger bowls with which genteel Europeans cleanse their hands after eating something like shellfish seem filthy by traditional Indian standards: the Indian custom is that hands should not be dipped into a basin of water, where they are recontaminated with the dirt they give off, but should be showered with a stream of fresh water for each person.

Those who dine with fingers are also very particular about *which* fingers they use. Not only is the left hand kept out of action (because it is used for toileting and therefore "unclean"), but there are strictures on which fingers of the right hand should be used. For true politeness, in most cultures where food is handheld, only the thumb, forefinger, and middle finger are used. (As with the various knife-and-fork rules, there are exceptions. Couscous, because it is so fragmented, may be eaten with all five fingers.) Food should not be grabbed precipitously from the common dish. It is also very rude to anticipate the next bite before you have finished the first, which is not the rule among knife-and-fork eaters.

As for the question of whether finger-eating limits what can be eaten, the answer is that it does, but no more so than forks or chopsticks. The main limitation is temperature. Cultures that eat with fingers do not have the same fetish for piping hot food and hot plates that we do. "Are your plates hot, Hot, HOT?" asked society hostess Elsie de Wolfe in 1934, in a guide to "successful dining." They'd better not be if you are eating with fingers. Room temperature, or a bit warmer, is the ideal for finger food. Fingers are also not the ideal tools for grappling with an English roast dinner: slabs of meat in gravy definitely call for cutlery.

In the countries of finger-eating, the food has evolved to fit, and hands have developed powers that the presence of cutlery denies them. Ottaviano Bon, a European traveler at the court of the "Turkish emperor" in the early seventeenth century, noted that the meat of the emperor was "so tender, and so delicately dressed that . . . he needs no knife, but pulls the flesh from the bones very easily with his fingers." Similarly, with a piece of Indian naan bread in one hand and a bowl of dal in the other, poised to dip and scoop, you do not feel the lack of a fork. Fingers are not just adequate substitutes for table utensils: they are better, in many respects. As Margaret Visser writes:

> To people who eat with their fingers, hands seem cleaner, warmer, more agile than cutlery. Hands are silent, sensitive to texture and temperature, and graceful—provided, of course, that they have been properly trained.

In Arab countries where eating with fingers is still the norm, people become dexterously nimble at manipulating food from hand to mouth. Many of the things that happen at mealtimes would be impossible with a fork: scooping up a ball of rice, then stuffing it with a piece of lamb or eggplant before popping it neatly into the mouth. No cutlery could improve on such a perfect and satisfying gesture.

The technology of tableware cannot be understood solely in terms of function. On pure utilitarian grounds, there is very little that you can do with the triumvirate of knife-fork-spoon or with chopsticks that you cannot do with fingers and a bowl (assuming that there is also some kind of cutting implement available). Table utensils are above all cultural objects, carrying with them a view of what food is and how we should conduct ourselves in relation to it. And then there are sporks.

The term *spork* is first recorded in a dictionary in 1909, though the first patent for one was only issued in 1970. Both the word and the thing are a hybrid of spoon and fork. Like a pencil with an eraser on the end, the spork is what theorists of technology call a "joined" tool: two inventions combined. In its classical form— fashioned from flimsy disposable plastic and given away at fast-food outlets—the spork has the scooping bowl of a spoon coupled

with the tines of a fork. It is not to be confused with a splayd, a knoon, a spife, or a knork.*

Sporks have developed an affectionate following, of a somewhat ironic kind, in our lifetime. There are several websites devoted to them, proffering tips on use ("bend the prongs inward and outward and stand the spork on end. This is a *leaning tower of spork*"), haikus in their honor ("The spork, true beauty / the tines, the bowl, the long stem / life now is complete"), and general musings. Spork.org has this to say:

> A spork is a perfect metaphor for human existence. It tries to func-
> tion as both spoon and fork, and because of this dual nature, it fails
> miserably at both. You cannot have soup with a spork; it is far too
> shallow. You cannot eat meat with a spork; the prongs are too small.

A spork is not one thing or another, but in-between. In the Pixar-animated film *Wall-E*, a robot in a postapocalyptic wasteland attempts to clear up the detritus left behind on planet earth by the human race. He heroically sorts old plastic cutlery into different compartments, until encountering a spork. His little brain cannot cope with this new object. Does it go with the spoons? Or the forks? The spork is uncategorizable.

Two years into his presidency, in 1995, Bill Clinton, pioneer of "Third Way" politics, made the spork the centerpiece of a humorous speech to the Radio and Television Correspondents' Dinner in Washington, DC. He claimed that the spork was "the symbol of my administration. . . . No more false choice between the left utensil

---

* Here's a quick translation: spork = a spoon with added tines; splayd = a knife, fork, and spoon in one, consisting of a tined spoon with a sharpened edge; knork = a fork with the cutting power of a knife; spife = a spoon with a knife on the end (an example would be the plastic green kiwi spoons sold in kitchenware shops); sporf = an all-purpose term for any hybrid of spoon, fork, and knife.

and the right utensil." He ended the speech, to rapturous applause and laughter: "This is a big, new idea—the spork!"

Clinton was being funny, but the spork in its way really is a big new idea. Where did it come from? An urban myth circulates that sporks were first invented by General Douglas MacArthur as part of the US occupation of Japan in the 1940s. The story goes that MacArthur decreed that chopsticks were barbarian tools, whereas forks were too dangerous (the fear being that the conquered Japanese might rise up and use them as weapons). Therefore, the spork was forced on the Japanese as a safe, truncated version of Western tableware. This story cannot be right—as mentioned above, the name *spork* dates back to before 1909 and the form itself is still older: in nineteenth-century American silverware, both terrapin forks and ice-cream spoons were sporks in all but name (they were also known as "runcible spoons" after the Edward Lear poem). It is true that as far back as World War I, various armies used foldable spoon-fork combinations in the mess kit; but these were not true sporks but rather a spoon and a fork riveted together at the handle. These utensils are still used in the Finnish military: they are made from stainless steel and called *Lusikkahaarukka*, meaning spoon-fork.

The urban myth about MacArthur and the Japanese possibly arose because the first person to create a hybrid fork-spoon for the mass market was another McArthur, an Australian named Bill McArthur, of Potts Point, New South Wales, who in 1943 launched his patented Splayd—derived from the verb to splay—after seeing a magazine photo of women awkwardly balancing knives, forks, and plates on their laps at a party. Boxes of stainless-steel Splayds— which described themselves as "all-in-one knife fork and spoon gracefully fashioned"—were marketed as the ideal solution to the newly popular Australian barbecue. They have since become an Aussie institution, having sold more than 5 million units.

In the 1970s, Splayds were finally joined by Sporks. The name was trademarked in 1970 by a US company (the Van Brode Milling Company) and in 1975 by a UK one (Plastico Ltd.) as a combination plastic eating utensil. It wasn't long before they became standard issue in fast-food restaurants. The spork made business sense: two plastic utensils for the price of one.

Other important users of the spork included schools and prisons and any other institutional setting where the business of feeding is reduced to its most basic, functional level. American prison sporks are generally plastic, orange in color, and very ineffectual, because it is vital that they should not be used as weapons. In 2008, a man was arrested in Anchorage, Alaska, for attempting an armed robbery with a spork from a fried-chicken restaurant. The victim's body was gashed with four "parallel scratches." The most remarkable thing about this story is that anyone could have managed to do such damage with a spork, which in its fast-food incarnation is a pitiful implement, splintering into plastic shards on contact with any remotely challenging food.

In 2006, the spork was given a radical reboot, which tried to address some of its structural shortcomings. Joachim Nordwall is a Swedish designer employed by the outdoor supplies company Light My Fire. Having grown up in Sweden, Nordwall had no background in using the fast-food spork, and he was not much impressed by it. "It feels like a compromise to me," he noted (to which one is tempted to say: duh!). The tines did not work well on their own terms as a fork. Nor did the bowl really work as a spoon: when eating soup, it would dribble out through the gaps. Nordwall's breakthrough was to separate out the spoon part and the fork part, placing them on opposite ends of the handle. For good measure, he added a blade to the outer edge of the tines, thus turning his construction into a sort of knork as well as a spork. "Sporks get a new look," raved a business review of Nordwall's design. Really, though, it was very old. Nordwall had reinvented the double-ended medieval sucket spoon.

There is now a spork for every occasion, except a meal where any degree of formality is required. Light My Fire sells brightly colored sporks for campers and sporks for office workers, "lefty" sporks for the left-handed, and "spork little" for toddlers. Unlike previous utensils, which always carried with them some cultural expectation of how you should behave in relation to food, the spork is entirely devoid of culture. It bends itself to the owner, rather than the other way around. It carries with it no particular mores and demands no etiquette. Eating with a spork is neither mannerly nor unmannerly. One of the many spork tributes on the Internet has fun with the notion of table manners for "sporkware," advising: "When using a spork to eat mashed potatoes out of a Styrofoam container, it is common courtesy to leave a little 'spork waste' at the bottom rather than scrape the styrofoam with the spork to get every last morsel. If you must have every little bit of potato, please use your finger."

～ *Tongs* ～

In the past, tongs tended to be specialist apparatus. Fire tongs for moving hot coals. Meat tongs for turning meat in the pan. Asparagus tongs for serving up the delicate green spears. Spring-loaded escargot tongs for gripping a slippery snail shell loaded with garlic butter.

Only now—since the 1990s—do we appreciate kitchen tongs for the versatile equipment that they are: an all-purpose lifting, prodding, and retrieval tool. I'm talking about the simple, cheap kind: stainless steel, scalloped at the edges, rather than the old-fashioned scissor tongs that mangle your food and snap shut when you are least expecting it.

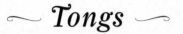

The function of tongs is to increase your dexterity at the stove. Holding tongs is like having heat-proof claws on the end of your arms. You become a creature capable of lifting up searing-hot roast chicken thighs or picking individual cardamom pods from a pilaf, with the accuracy of tweezers and the calmness of a spatula.

Tongs are best on the short side (8 inches is ideal). The longer the tongs, the trickier they are to manipulate, which defeats the object. Classically trained French chefs once used long, bone-handled two-pronged forks to do the same jobs. But a fork is more limited in its scope. It cannot pull linguini out of a pot of boiling water at the very moment it is done, before deftly tossing it with ham, peas, and cream. With tongs at your disposal, you technically need neither colander nor pasta servers.

Other than a knife and a wooden spoon, they are the most useful handheld utensil I know.

⁊[7]⁊

ICE

I have eaten
the plums
that were in
the icebox
... Forgive me
they were delicious
so sweet
and so cold

WILLIAM CARLOS WILLIAMS,
"This Is Just to Say"

J ULY 24, 1959, WAS A VITAL MOMENT IN THE COLD WAR. Nikita Khrushchev, the leader of the Soviet Union, and Richard Nixon, then vice president of the United States during Eisenhower's presidency, staged a big public meeting in Moscow, in front of TV cameras. It was the most high-profile exchange between Soviets and Americans since the Geneva Summit of 1955, but far more informal. Laughing and sometimes jabbing fingers at one another, the two men debated the merits of capitalism and communism. Which country had the more advanced technologies? Which way of life was better? The conversation—which has since been christened the Kitchen Debate—hinged not on weapons or the space race but on washing machines and kitchen gadgets.

The occasion was the opening day of the American National Exhibition at Sokolniki Park, a municipal park of "leisure and culture." This was the first time many Russians had encountered the American lifestyle firsthand: the first time they tasted Pepsi-Cola or set eyes on big American refrigerators. The exhibition featured three fully equipped model kitchens. One was a labor-saving kitchen from General Mills, with an emphasis on frozen foods. Another was a "futuristic" kitchen made by Whirlpool that would require women only to push a button to set off all sorts of kitchen machines. The third was a lemon-yellow modular kitchen supplied by General Electric.

This third kitchen is the one that entered the history books. It looked like something out of a Doris Day movie: clean and yellow and apple-pie neat. Pretty female demonstrators showed Russian visitors the wonders that could emerge from the lemon-yellow fridge: cupcakes topped with cool creamy frosting, fudgy chocolate layer cake. This General Electric kitchen was made as part of a complete ranch-style American house.

Nixon and Khrushchev paused to look at the kitchen, leaning against the white dividing railings. Lois Epstein, a perky brunette American exhibition guide, demonstrated how the typical American housewife might use the built-in washer-dryer. On top of the machine were a box of SOS scouring pads and a box of Dash soap powder. "In America, we like to make life easier for women," noted Nixon. Khrushchev replied that "your capitalistic attitude to women does not occur under communism," implying that instead of making life easier, these machines only confirmed the American view that the vocation of women was to be housewives (and perhaps he was partly right about this). Khrushchev went on to query whether all these new machines offered real benefits. In his memoirs, Khrushchev recalled how he picked up an automatic device for squeezing lemon juice for tea and said, "What a silly thing . . . Mr. Nixon! . . . I think it would take a housewife longer to use this gadget than it would for her to . . . slice a piece of lemon, drop it into a glass of tea, then squeeze a few drops."

Nixon fought back, drawing Khrushchev's attention to all the gleaming utensils on display—mixers, juicers, can openers, freezers. "The American system," he insisted "is designed to take advantage of new inventions." Khrushchev continued to be disdainful. "Don't you have a machine that puts food in the mouth and presses it down? Many things you've shown us are interesting but they are not needed in life. They have no useful purpose. They are merely gadgets."

Yet Khrushchev wanted to have it both ways. While rejecting the American kitchen as worthless, he also wanted to insist that the Soviets could make kitchens that were every bit as good. He wanted to win the kitchen race as well as the space race. "You think the Russian people will be dumbfounded to see these things, but the fact is that newly built Russian houses have all this equipment right now." This wasn't true, and Khrushchev surely knew it. No house or apartment in the whole of Moscow had a kitchen anything like the gleaming yellow exemplar at the American exhibition. By American standards, Soviet kitchens in the brand-new apartments built under Khrushchev's regime were miniscule: between 4.5 and 6 meters square. The crowning glory of these kitchens, the great labor-saving device of the future was a series of cramped wall-mounted cabinets and cupboards underneath the work surfaces. These were built at a standard height—85 cm—aimed at Mrs. Moscow Average. Taller women must stoop; smaller ones must stretch, to bend themselves to the state's uniform standards. Apart from work space, what these kitchens notably lacked was anything like the capacious lemon-yellow refrigerator in the General Electric model kitchen. Soviet fridges in 1959 were ugly and cramped. But the vast majority of Russian kitchens had no fridge at all.

The fact was that neither the Soviet Union—nor any country in the world, not Britain, not Germany—could match American domestic refrigerators in 1959. The United States was the ice nation par excellence. Ninety-six percent of households owned fridges (compared to 13 percent for Britain). The American way of life was, to a very large extent, made possible by refrigeration. From the clink of ice cubes in a glass of bourbon to the easy luxury of a Chicago steak in New York City, from soda fountains and popsicles to frozen peas, the business of cooling food and drink was deeply American. An automatic lemon squeezer may indeed be, in Khrushchev's words, a "mere

gadget"; a refrigerator is something more. It serves many useful purposes, being not just a single technology but a cluster of interrelated technologies that together created entirely new approaches to eating. Sometimes, refrigeration is a tool for making something cold for the sheer frivolous delight of it—an icy glass of white wine, a refreshing slice of melon. But it is also a method of food conservation: keeping food safe to eat for longer times and over longer distances. The efficient home refrigerator entirely changed the way food—getting it, cooking it, eating it—fitted into peoples' lives.

The roomy American fridge—and its close relative, the freezer—was, first and foremost, a tool for food preservation, which freed cooks from the need to pickle, salt, or can that which could not be eaten straight away. At a stroke, it removed the harsh seasonality of consumption, for the poor as well as the rich. The fridge transformed what people ate: fresh meat, fresh milk, and fresh green vegetables became year-round staples in all parts of the United States for the first time in history. Fridges changed how people shopped for food: without refrigeration, there could be no supermarkets, no "weekly shopping," no stocking up the freezer for emergencies. As well as a preservation device, the fridge was a storage system, taking over the function of the old larder. To have a refrigerator heaving with fresh produce—lettuces in the crisper, gallons of milk, jars of mayonnaise, whole roast chickens, pounds of cold cuts, and creamy desserts—was to participate in the American Dream, which is at heart a dream about plenty. The American fridge became a new focal point for the kitchen, taking over from the old hearth. Once, we congregated around fire; now people organize their lives around the hard, chilly lines of the refrigerator.

The whole world aspires to be American now, when it comes to fridges. In spring 2011, at a product launch in a vast underground space in London's Bloomsbury, I stood in front of a new state-of-the-art refrigerator-freezer. Its eco rating was A++ and it was frost free. It was tall and plain white, except for a little metallic panel on the front, which looked like a security device from the world of James Bond. There was a button you could press with a parasol on it if you went on holiday: for all the time you were away sunning yourself, the fridge would calibrate its energy use to a lower setting than normal. I was impressed. But this was nothing. Samsung has launched a "smart fridge," with built-in Wi-Fi, Twitter feeds, and weather reports. At the time of this writing, researchers at the University of Central Lancashire in Britain were working on a self-cleaning fridge that would also do a constant inventory of its own contents, moving goods nearing their use-by-date toward the front. It seems we have reached the point where we expect a fridge to organize our lives for us; and where they will soon be capable of doing so.

A fridge rather than a stove now tends to be the starting point—what designers call the "statement"—around which the rest of a kitchen is constructed. When we can't think what else to do, we open the fridge door and stare into it long and hard as if it will provide the answers to life's great questions.

Bacon, parmesan, and cheddar, hard sticks of salami, sauerkraut, confit duck, sausages, smoked salmon, kippers, salt cod, sardines in oil, raisins, prunes, dried apricots, raspberry jam, marmalade. . . .

Countless delicious foodstuffs might never have been invented if refrigeration had been available sooner.

The continued presence of many of these foods in our diet is an anachronism; but we are creatures of habit and have learned to love many things that were once eaten from necessity. Bacon serves no real purpose in a refrigerated age, except that of pleasure, which can never be discounted. There is no longer any need to eat smoked ham when you can keep a fresh pork chop in the fridge. Our taste for smoked things belongs to earlier times, when preserving meats by smoking them could make the difference between being able to eat a food year-round and eating it just once a year.

For the whole of winter and spring in medieval times in Europe, almost all protein foods—if you were lucky enough to have them— would have been smoky and salty because this was the only way to stop meat and fish from spoiling. Any meat not eaten immediately after slaughter would be preserved by salting: portions of meat were layered in a large wooden cask, smothered in layers of salt. This was an expensive process: as of the late thirteenth century, it took two pennies' worth of salt to cure five pennies' worth of meat, so only good-quality meat was salted. Pork was the meat that took salt the best. As well as ham and gammon, bacon and salt pork, the Elizabethans made something called, crudely, "souse," a pickled medley of pigs' feet, ears, cheek, and snout: everything but the squeak. Salt beef was also made. A variant was "Martinmas beef," prepared around the feast of Martinmas, on November 11. After it was salted, the meat was hung in the roof of a smoky house until it was well and truly smoked. For a long time, the myth was bandied about that cooks in the past used spices to disguise the taste of putrid meat. This was not so: spices were expensive and would not be wasted on condemned food. But an important use of spices was tempering the harshness of salt meat.

Perishable milk was preserved, too: in the east, by curdling and fermenting it into yogurty products and sour drinks such as the Kazakh *kumiss*, or by evaporating it into milk powder (a Mongol invention); and in the West by turning it into highly salted cheese or butter, packed in well-glazed earthenware pots. In Aelfric's *Colloquy* the "salter" remarks, "You would lose all your butter and cheese were I not at hand to protect it for you." Medieval salty butter was far saltier than our "salted butter," which is seasoned to suit our palate rather than for preservation. A typical modern salted butter contains 1–2 percent salt, whereas medieval butter contained 5–10 times as much: according to a record of 1305, 1 pound of salt was needed for 10 pounds of butter, i.e., the butter was 10 percent salt. Eaten straight, this would have been foul. Cooks needed to go to great lengths to wash much of the salt out again before it was used.

Salt was also used to preserve the fragile flesh of fish. The Scottish kipper was not invented until the nineteenth century, but before that, there were Smokies and Buckies and Bervies, all names for cured haddock produced near Aberdeen, heavily smoked over peat and decayed moss. Salted and pickled fish were staple proteins in Europe, particularly on Fridays. Since preclassical times there had been a significant trade in salted fish, first from Egypt and Spain, then from Greece and Rome. During the Middle Ages, making salt herring from the North and Baltic Seas was a major industry. It was not an easy product to manufacture. As an oily fish, herring turns rancid very fast and ideally it should be salted within twenty-four hours of being caught, or even sooner. In the fourteenth century, herring merchants were able to streamline the process considerably once they developed techniques for salting the herring on board ship; the fish was then repacked once they reached the shore. The Dutch in particular proved masters at this, which may be how they achieved their dominance of the European market. Dutch herring gutters could process up to 2,000 fish an hour at sea. This speed had an additional benefit, though the gutters would not have been aware of it. In their haste, they failed to eviscerate a part of the stomach containing trypsin, a chemical that speeds up the curing process.

The monotony of a diet in which the only fish you ever ate was preserved may be gauged by the number of jokes these items gave rise to. "You dried stockefish, you, out of my sight!" says one character to another in *A Pleasant Comedie, Called Wily Beguilde* (Anon., 1606). A red herring—which was a particularly pungent cured fish, double "hard smoked" as well as salted—remains in our language as something comically deceptive or out of place.

Sweet preserved foods tended to have much more luxurious and pleasurable connotations. In hot Mediterranean countries, the most expedient way to preserve fruits and vegetables was to dry them: grapes became "raisins of the sun," plums became prunes, dates and figs shriveled up and intensified in sweetness. The basic technology of drying fruit was very simple: in biblical times and before, juicy fruits and vegetables were buried in hot sand or spread out on trays or rooftops, to become desiccated in the sun's rays. In Eastern Europe, however, where the sun was less fierce, they developed more sophisticated methods. Beginning in the Middle Ages, in Moravia and Slovakia, special drying-houses were built, with rooms filled with wicker frames laden with the prepared fruit, with constantly burning stoves set underneath the frames to generate enough dry heat to convert apricots into long-lasting dried apricots and cherries into chewy dried cherries.

The equivalent in the rich houses of England was the cool "still-room," in which servants distilled spirits, bottled fruits, candied nuts and citrus peel, and made marmalades (originally, from quinces), jams, and sweetmeats. The art of candying was rife with alchemical superstitions and "secrets." Each fruit had its own imperatives. According to a medieval book, walnuts should be preserved on June 24, St. John's Day. Fruits for preserving were picked almost ripe rather than fully ripe, because they were more likely to hold their shape. "The best way to preserve gooseberries green and whole" was a recipe in Hannah Wolley's *Queen-like Closet* of 1672. Wolley's method was tediously long: three soakings in warm water, three boilings in sugar syrup, then a final boiling in a fresh sugar syrup. No wonder it was hard work: the work of the stillroom was a kind of magic, a staving off of decay comparable to the embalming of the dead.

The most remarkable thing about fruit preserves was the fact that they really did preserve the fruit (at least, most of the time). Throughout history, cooks have aimed to make food safe to eat; and often, they succeeded. Yet until the 1860s, when Louis Pasteur uncovered the microorganisms responsible for spoiling food and drink, cooks had no real knowledge of why food preservation worked. The prevailing view was that decomposition was caused by spontaneous generation, in other words, that mysterious unseen forces caused the mold to grow. People knew nothing of  microbes, the living organisms—fungi, bacteria, and yeasts, among others—that cause beneficial fermentation in wine and cheese, and toxic fermentation when food degrades.

Greek women spreading figs in the sun to dry did not know that they were killing off invisible microbes (bacteria need moisture to flourish and when food dehydrates, they mostly die). The farmers' wives who pickled onions in vinegar did not understand how acidity protects against the growth of molds (microbes prefer alkaline conditions)—they just knew that pickled onions kept longer than unpickled onions. Methods of preservation developed slowly and cautiously. Keeping food safe to eat was a process of trial and error; but because error could mean death, there was little incentive to embark on new trials. Having found something that did successfully keep food edible for a long period of time, you stuck to it. Except for the sixteenth-century discovery of conserving meat in a layer of fat or oil (whether duck confit or the potted meats of England), the technology of preservation made no advances from the Middle Ages to the beginning of the nineteenth century. Then came the can.

Even Nicolas Appert, the Frenchman who invented canning—a very modern system of preservation—did not fully comprehend how or why it worked. He claimed it was "the fruit of my dreams, of my reflections, of my researches." Appert was originally a brewer who next worked as a steward to the aristocracy, and later, in Napoleonic times, as a confectioner. He was said to be a cheerful individual, with very thick black eyebrows and a bald head. Although he made one of the great advances in food technology of the nineteenth century, he derived no lasting benefit from it and was buried in a pauper's grave.

In 1795, the French government, then embroiled in war with Britain, sought better ways of feeding its armed forces. Napoleon offered 12,000 francs to anyone who could come up with the best new way of preserving food. Meanwhile, Appert, then running his confectionery on the rue des Lombards in Paris, was preoccupied with the same question. He knew how to conserve and candy countless different fruits in sugar, but he felt sure that there must be a more "natural" way of achieving the same effect. In Appert's view, all of the traditional methods of preserving were faulty. Drying foods took away their essential texture; salt made foods "acerbic"; sugar concealed the true flavors. Appert sought a technique that would preserve without destroying the true characteristics of any given ingredient. He experimented with conserving fruits and vegetables and meat stews in champagne bottles, heating them in baths of hot water. Over time, he switched the champagne bottles for wider-necked bottles. Eventually, he felt confident enough to send a few samples to the French navy. The response was positive: the minister of the navy commented that Appert's beans and green peas had "all the freshness and flavor of freshly picked vegetables." *Le Courrier de l'Europe* was still more fulsome in its praise: "M. Appert has found a way to fix the seasons." The 12,000 franc prize was duly awarded to him.

Appert's method was very simple. It consisted of nothing more than heating food in a water bath in corked bottles. In 1810, Appert published a book revealing his secrets. The foods that Appert preserved in his corked bottles were exotic: artichokes, truffles, chestnuts, young partridges, grape must, sorrel, asparagus, apricots, red currants, soup of julienned vegetables, new-laid eggs. But in essence it was the same process by which every can of tuna and every can of corn kernels is still manufactured: in a sealed container heated in steam.

Appert, however, was not the one to capitalize on it. By accepting his prize, he forwent the chance to patent his invention. Just months after Appert's book on canning came out in 1810, an English broker, Peter Durand, rushed out a patent for a method of food preservation suspiciously similar to Appert's. The patent was purchased for £1,000 by Brian Donkin, an engineer with an eye for the main chance. In 1813, Donkin, with his business partners, Messrs. Hall and Gamble, opened a factory nicknamed the "Preservatory" in Bermondsey, churning out foods processed by Appert's technique, heating them in closed containers in boiling water for as long as six hours. There was one crucial difference. They found Appert's glass bottles to be too fragile. Instead, Donkin, Hall, and Gamble packed their food—carrots, veal, meat soup, boiled beef, and suchlike—into tin-coated iron canisters: tin cans.

These early cans of food were not without problems. The most immediate was that there was a lag of fifty years between Appert's discovery and the invention of the first can openers. This is a glaring example of how technology can proceed in fits and starts. Until the 1860s, cans of bully beef (much used by armies) or salmon or cling peaches would come with instructions to "cut round the top near the outer edge with a chisel and hammer."

The first custom-made can opener was designed by Robert Yeates, a maker of surgical instruments and cutlery, in 1855: it was a vicious clawlike lever attached to a wooden handle. The idea was to gouge the lever into the top of the can and then forcefully cut around, leaving a jagged edge. It did the job—but not well. The history of can openers is riddled with unsatisfactory designs: the Warner, much used during the American Civil War, with a sharp sickle on the end, fine for a battlefield but deadly for use in a normal kitchen; an 1868 key to open a can by rolling a strip of the top metal off, which turned out to be ideal for sardine cans, not so good for the normal cylindrical tins, because it only opened a portion of the circular lid; the electric openers of the 1930s that introduced an unnecessary element of complexity to the task at hand. Finally, in the 1980s, a device appeared that did the job with minimal danger or exertion for the user. The side-opening can opener—which can now be purchased at modest cost in many variations—is one of the great unsung heroes of the modern kitchen. Instead of piercing the top of the can, it uses two wheels in tandem, one rotating, one serrated, removing the entire lid and leaving no sharp edge. It's an inspired tool, the only shame being that it wasn't invented sooner. The canned food industry is now shifting to pop-top self-opening cans, eliminating the need to own a can opener at all.

Aside from the challenge of getting at the food inside the cans, canning posed another danger: it did not always succeed in preserving the food. In 1852, thousands of cans of meat supplied to the British navy were inspected and found to be unfit to eat, "their contents being masses of putrefaction" causing a dreadful "stench" when opened. It was generally assumed that canned meat spoiled because "air has penetrated into the canister, or was not originally entirely exhausted." Until Louis Pasteur, it wasn't known that there is a class of microbe that can flourish without air: to kill these, the crucial factor is thorough heating. The original size of cans had been around 2 to 4 pounds (as against ¼ to 1 pound for average cans today); these navy cans were massive, holding on average 10 pounds of meat. The heating time in the factory should have been correspondingly increased, but it wasn't, leaving putrid pockets in the middle of the can.

By the 1870s, the quality of canned food had improved, and cans were starting to open up global food markets as never before. British workers sat down to a supper of Fray Bentos corned beef from Uruguay. Canned hams traveled all the way from Bermondsey to China. American consumers were introduced to ingredients they might seldom otherwise have tasted. One historian of canning noted that the American family could now pick from "a kitchen garden where all good things grow . . . filled with raspberries, apricots, olives, and pineapples," not to mention "baked beans."

It was a garden, however, in which many of the plants tasted slightly strange. True, Italian canned tomatoes can be a joy—not by themselves, but slow-simmered in countless pasta sauces: puttanesca, amatriciana. But spinach in a can—sorry, Popeye—is sludgy and metallic. Canned pineapple and peaches are fine (though they lack the perfume of fresh fruit), but canned raspberries are mush. Today, cans are more significant as packaging for drinks (fizzy sodas, beer) than food: world sales of processed food in cans are around 75 billion units a year, as against 320 billion units for canned drinks.

In the end, the technology that most improved the diet of American families was not canning but refrigeration, which really did give people access to "a kitchen garden where all good things grow."

In 1833, a surprising consignment arrived in Calcutta, then the center of the British Empire in India. It was forty tons of pure crystalline ice, which had come all the way from Boston on the East Coast of the United States, a journey of 16,000 miles, shipped by Frederick Tudor, an ice entrepreneur.

The Boston-to-Calcutta ice trade was a sign of how America was turning ice into profit. As an abundant natural resource, ice is ancient. There were ice harvests in China before the first millennium BC. Snow was sold in Athens beginning in the fifth century BC. Aristocrats of the seventeenth century spooned desserts from ice bowls, drank wine chilled with snow, and even ate iced creams and water ices. Yet it was only in the nineteenth century in the United States that ice became an industrial commodity. And it was only the Americans who recognized and exploited the fact that the biggest bucks were not in making icy treats but in using ice for refrigeration: the preservation of food.

Cold storage was not unknown before the nineteenth century. Many estates in Italy had their own icehouses, such as the one in the Boboli Gardens in Florence. These were pits or vaults, heavily insulated—usually with turf or straw—in which unevenly hacked slabs of winter ice could be kept cold for the summer. These houses were not, however, principally devices for preserving food but were for preserving ice, so that it would be ready for cooling drinks or making lavish ice creams at the height of summer. The icehouse may sometimes have been used to supplement a larder—but the primary function was keeping its owners supplied with sweet cold treats, the accoutrements of civilized living. To have access to ice in the summer—to flout the seasons—was a sure sign of wealth. "The rich get their ice in the summer, but the poor get theirs in the winter," as Laura Ingalls Wilder wrote in her book about life married to a struggling farmer on the Dakota prairie in the 1880s.

In America at large—a country of vast distances and extremes of climate—the lack of ice affected the entire food supply. Butter, fish, milk, and meat could only be sold locally. Most butchers killed only what meat they could sell in a single day. Unsold meat—known as the shambles—was left to rot on the streets. Unless you were a country dweller with a kitchen garden, green vegetables were a rarity. The basic diet was salt pork and bread or corn bread. Consumers in the city and producers in the countryside had few ways to reach one another. In 1803, an enterprising Maryland farmer named Thomas Moore worked out that he could sell more butter if he could take it further to market. Moore created one of the very first "refrigerators": an egg-shaped cedar-wood tub, with an inner metal container for the butter. Between the metal and the wood, there was a gap, which could be packed with ice.

The first great technological breakthrough in the American ice industry was the horse-drawn ice cutter, patented by Nathaniel J. Wyeth in 1829. Before this, ice was harvested—with great difficulty, using axes and saws—in uneven blocks. With far less effort, for the humans, if not the horses, Wyeth's ice cutter produced perfect square blocks, easy to stack and transport. There were epic profits to be made. As of 1873, it cost 20 cents a ton to harvest ice on the Hudson River. This could be sold on to private customers for as much as $4 to $8 a ton, a potential profit margin of 4,000 percent.

In 1855, horsepower was joined by steam power in the ice harvest, and as many as 600 tons could now be harvested in a single hour. Supply increased; but so did demand. In 1856, New York City used 100,000 tons of ice; in 1879–1880, the city needed nearly 1 million tons, and rising. Nearly half of all the ice sold went to private families. Ice companies delivered ice on wagons or trucks for a flat daily or monthly charge. Ice was kept in an icebox—a primitive refrigerator, little more than a tin- or zinc-lined wooden box with shelves like a kitchen cabinet, with a drainage hole at the bottom for the meltwater. Iceboxes were smelly and inefficient, with no means of circulating the air. But still—what a boon—to be able to enjoy cold-ish food on a July day; to stop fresh milk from going sour for a few hours, if not days; to chill a bowl of plums.

Ice wrought its greatest nineteenth-century transformations, however, not in private homes but in the commercial food supply. A combination of vast cold-storage warehouses and refrigerated railroad cars opened up entirely new food markets. The meat, dairy, and fresh produce industries were the biggest winners. By the time of World War II, Americans were known around the world for their seemingly inordinate appetite for meat and milk (supplemented with glasses of freshly squeezed orange juice and green salads). This appetite—and the means of satisfying it—was largely a creation of nineteenth-century refrigeration.

In 1851, butter was first transported in refrigerated railroad cars from New York City to Boston. Fish, too, began to travel the country, and in 1857, fresh meat went from New York to the western states. Refrigerated "beef cars" created a new meatpacking industry, centered in Chicago. This was a very American phenomenon: by 1910, there were 85,000 refrigerated cars in the United States, compared to just 1,085 in Europe (mostly in Russia). Fresh meat no longer had to be slaughtered and used immediately. "Dressed beef" could be cooled, stored, and shipped anywhere.

The new refrigerated cars had fierce critics, as do all new food technologies. Local butchers and slaughterhouses objected to the loss of business and lamented Chicago's growing monopoly on meat (and judging from the horrific conditions in Chicago meatpacking factories described in Upton Sinclair's *The Jungle*, they may have had a point). More generally, the population at large was scared of the very thing that made refrigeration so useful: its ability to extend the storage time of food. Alongside the growth in refrigerated cars was a huge growth in cold-storage warehouses. By 1915, 100 million tons of butter in America were in cold storage. Critics argued that "delayed storage" could not be good for the food, reducing its palatability and nutritional value. Another persistent worry was that cold storage was a scam: by delaying the sale of produce, the sellers could push prices up.

One more concern about refrigeration—particularly for dairy foods, whose storage needs to be scrupulously clean—was that natural ice was not always pristine, often containing dirt, pond weeds, and other vegetation. Local boards of health would periodically condemn large amounts of naturally harvested ice as unfit for human consumption.

This was one of the reasons refrigeration in America increasingly moved from natural ice to factory-made ice. Humans had known ways of making artificial ice for centuries, but by and large they had done so not for the purposes of refrigeration but for making ice creams and cold drinks. The Elizabethan scientist Francis Bacon was one of the few exceptions. According to the biographer John Aubrey, Bacon died in 1626 from a chill contracted while attempting to use snow to preserve a chicken. He also conducted investigations into the use of saltpeter, for what he called "the experiment of the artificial turning of water into ice." Bacon attacked the frivolous uses to which the rich tended to put their ice. It was "poore and contemptible" to manufacture ice purely for such niceties as cooling their wines instead of using it for "conservatories," by which he meant refrigerators. Bacon was surely right that this was largely a question of priorities. Whereas refrigeration was neglected for centuries, the technology of ice cream was extremely advanced.

The ad for Mrs. Marshall's Patent Freezer, an ice-cream maker from 1885, shows a picture of a shallow, circular, hand-cranked machine and includes this boast:

MARSHALL'S PATENT FREEZER
. . .
Smooth and Delicious Ice produced in three minutes.

Ice cream in three minutes? By hand? Today, the top-of-the-line electric ice-cream makers aimed at home cooks at a cost of $100 to $200 boast that they can deliver "ice cream or sorbet in less than thirty minutes." How could Mrs. Marshall's machine possibly have made ice cream in a tenth the time, without the aid of electricity?

It sounds like commercial hype. Mrs. Marshall was an extremely shrewd businesswoman, skilled at promoting her own interests. A mother of four from St. John's Wood in North London, she

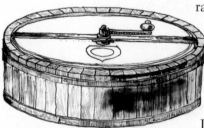

ran a school of cookery at 31 Mortimer Street, first established in 1883. Judging from the portraits in her books, she was an attractive brunette, rather in the mold of the dark beauties painted by John Singer Sargent: a bright gaze, bosomy blouses, tumbling curls piled up on her head. Fairly soon after her cookery school was established, Mrs. Marshall branched out with a shop, offering to equip entire kitchens with every utensil and appliance necessary, from knife cleaners to ornate ice-cream molds. She also sold essences, condiments, and food colorings and wrote cookbooks—two on ice cream and one on general cookery—always including plenty of ads at the back of the book for her own range of products.

In short, Mrs. Marshall comes across as exactly the sort of person who might claim her ice-cream maker took three minutes when really it took thirty. Sometimes, however, self-promoters actually do have something to boast about. As it transpires, Marshall's Patent Freezer truly is a spectacular device. As of 1998, there were only five known machines remaining in existence. Three of them were owned by Robin Weir, Britain's leading historian of ice cream and one of a small but passionate group of food historians who argue that Mrs. Marshall was a far greater cook than her near-contemporary Mrs. Beeton. When Weir started to experiment with his original Marshall's Patent Freezers, he was taken aback to find that they really could produce soft, creamy ice cream in just a few minutes—if not quite three, then no more than five, assuming the batch was not too large.

I have seen a Mrs. Marshall machine in operation during one of Ivan Day's historic cookery courses (Day is one of the other rare people to own one, and another Mrs. Marshall champion). To look at, it doesn't seem so very different from the classic American hand-cranking machine invented in 1843 by Nancy Johnson, the wife of a US naval officer from Philadelphia, another great female ice-cream innovator. These homespun wooden Johnson-style buckets are still brought out in many American households as a way of keeping kids amused on a hot summer's afternoon. Ice and salt are packed into the bucket around a metal container. Then the ice-cream mixture is poured into the container. The lid of the container is replaced, and you start to crank the handle, which turns the "dasher" inside, scraping the ice cream from the cold sides of the container as it freezes. On a good day, when conditions are not too warm and you've managed to pack in the maximum amount of ice and salt, the ice cream will be ready after twenty minutes of vigorous cranking.

Mrs. Marshall's Patent Freezer does the same job four times as fast. How?

It is much wider and shallower than the Nancy Johnson bucket design. Freezing is a form of reverse heat-transfer. Heat flows away from the custard mixture to the chilly metal container. The greater the surface area of the cold metal, the quicker the ice-cream mixture freezes. Mrs. Marshall's freezer has a much larger cold surface than other ice-cream makers. Unlike in Johnson's bucket, the ice and salt only go under the pan. As the ad boasts, "There is no need to pack any round the sides." It has another innovation, too. In every other domestic ice-cream machine in existence, whether electrical or pre-electrical, the metal container stays still while the paddle moves around. In Mrs. Marshall's freezer, the central paddle stays still, while a crank in the top turns the metal container around and around.

It is a superb invention, with just one flaw. In order to make it as affordable as possible, Mrs. Marshall manufactured her machines from cheap zinc, a poisonous metal. Therefore, although the remaining machines in existence undoubtedly do a great job of making silky-smooth gelato in a very short time, no one has tasted any in a very long time, except for Robin Weir. He tells me that he "eats ice cream out of mine all the time; at subzero temperatures the toxicity of metals becomes negligible." No doubt he is right about this, but in today's world any machine that toxifies your ice cream with zinc, however mildly, is not going to find many users.

At Ivan Day's house, we watched as a batch of citrus- and bergamot-scented water ice turned from a translucent yellow liquid to a snowy white cream. The temptation to taste some was immense, poisonous or not. Day remarked that he and Robin Weir have spoken of relaunching the Mrs. Marshall machine in modern nontoxic materials. They should. It is better than anything now on the market: quicker, more efficient, more aesthetically pleasing, and entirely carbon-neutral to operate. For someone in possession of a Marshall's Patent Freezer, it was probably easier and quicker to make home-made ice cream in 1885 than it is in most kitchens now. Even the revolutionary new Pacojet, which claims to make frozen desserts in twenty seconds by "precision spinning," is actually slower than the Marshall freezer: when "pacotizing," you need to freeze ingredients for at least twenty-four hours before you start. What's still more

noteworthy about Mrs. Marshall's innovation is that ice-cream making is not some neglected art (jelly molds were better on average a
hundred years ago than they are today, but that's because most of us
aren't that interested any more in building jellies to look like
palaces). Many cooks now would love to be able to do what Mrs.
Marshall managed. The range of ice-cream flavors in her *Book of Ices*
speaks of the freedom she had to invent anything she liked, in the
knowledge that once the mix was made, the finished ice could be
ready in minutes: she has recipes not just for vanilla, strawberry, and
chocolate but also for burnt almond, gooseberry, greengage, cinnamon, apricot, pistachio, quince, orange flower water, tea, tangerine.

Mrs. Marshall thought up another astonishing ice-cream idea. In
an article in her magazine, *The Table*, from 1901, she suggested an
amusing trick for "persons scientifically inclined":

> By the aid of liquid oxygen . . . each guest at a dinner party may
> make his or her own ice cream at the table by simply stirring with a
> spoon the ingredients of ice cream to which a few drops of liquid
> air has been added by the servant.

She probably got the idea from seeing a scientific lecture on liq-
uefied gases at the Royal Institution. It is not clear if she ever actually tried it herself. The scientist Peter Barham—who makes ice
cream using liquid nitrogen—suggests not, because "a few drops" of
liquid oxygen would probably not be enough to freeze a whole bowl
of ice cream.

Nevertheless it is striking that, with the twentieth century just
dawning, this great culinary innovator had devised a method of making ice cream that would still seem high-tech over a hundred years
later. Diners at Heston Blumenthal's The Fat Duck, probably the
most cutting-edge restaurant in England, still gasp when they see
desserts frozen at the table with liquid nitrogen.

Mrs. Marshall's liquid air came at the end of hundreds of years
of ice-cream innovation. The basic device of adding salt to ice to

lower its temperature was discovered around 300 AD in India. It works because salt lowers the freezing point of ice—in theory as low as −5.8°F. By the thirteenth century, Arab physicians were making artificial snow and ice by adding saltpeter to water, preempting Bacon by more than three centuries. Visitors from Europe to the East were struck by the wonderful sherbets and chilled syrups. Pierre Belon was a Frenchman who visited the Middle East in the sixteenth century. He marveled at the sweet cold drinks:

> Some are made of figs, others of plums, and of pears and peaches, others again of apricots and grapes, yet others of honey, and the sherbert-maker mixes snow or ice with them, to cool them.

In Persia, sherbets were made from lemon, orange, or pomegranate juice. First, the fruit was squeezed through a silver strainer. Sugar was added, and water to dilute. Finally, ice was piled in. Like the icy gola drinks still made on the beaches of India, this was somewhere between lemonade and a slush puppy: a cooling balm on a sweltering afternoon. "Give me a sun, I care not how hot," wrote the poet Byron when he visited Istanbul in 1813, "and sherbet, I care not how cool, and my Heaven is as easily made as your Persian's."

By the seventeenth century, Europeans were making their own frozen water ices in Paris, Florence, and Naples, and by the mid-eighteenth century, sweet ices were a well-established food. *Sorbetto* sellers walked the streets of Naples (*sorbetto* rather than *gelato* was the all-purpose Italian term for ice cream at this time, and did not signify a lack of cream), offering ice cream for sale with flavors including sweet orange, bitter cherry, jasmine, and muscat pear. It was dolloped out of the *sorbettiera* in which it was made. This was a tall, cylindrical metal container with a lid set in a bucket of ice and salt. To break up the ice crystals and keep the *sorbetto* creamy as it froze, the sellers spun the *sorbettiera* around on the ice-salt mixture every few minutes, an action that churned the mixture inside. Once in a while, the ice would be stirred with a wooden spatula. This is another low-tech way of making ice cream that can produce results every bit as good as our giant electric machines.

In short, we have very little to teach our ancestors when it comes to domestic ice-cream making. Our main nonelectric method for making sorbet—freezing it in a plastic container in the freezer, whisking every so often to break down the ice crystals—is hopelessly inferior to either a *sorbettiera* or a Marshall's Patent Freezer. No matter how often you take it out and whisk, the end result is always an unappealing slab of ice. Aside from industrial ice-cream manufacture—which in most cases is the art of cheapening the product with air and additives—there have been few real innovations since Mrs. Marshall's day.

Given the Victorian mastery of ice-cream technology, it might be expected that refrigeration was the obvious next step. Certainly, in the grandest houses of Europe, where the kitchen servants were divided up into savory cooks and sweet confectioners, the confectioners sometimes had access to a "cold room" where pastry could be kept cool, ices made, and meat stored. In more modest houses, however, refrigeration was still in its infancy, long after the Industrial Revolution. In the 1880s, Mrs. Marshall sold a range of "Cabinet Refrigerators" with "all modern improvements." These were nothing more than freestanding wooden kitchen cabinets with a couple of containers for ice in the top. Whereas Marshall's Patent Freezer is one of the great neglected kitchen technologies, her fridges were nothing but Victorian curios. They have been rendered entirely obsolete by the electric compressor fridge that now shapes all our lives.

A few years ago, I was talking to an American in London who was feeling homesick. The thing that really got her, she said, was that her British kitchen, with its poky little appliances, was too quiet. She missed the hum—not loud, but constant—that comes with a big American fridge. For her, this hum was the sound of home.

It was not inevitable that twentieth-century American fridges would develop this friendly hum, which is a consequence of the motor inside electric refrigerators (large fridge = large motor = loud hum). There was another technology available that was potentially no worse than the electric fridge: the gas absorption fridge, whose operation was silent. Both methods of refrigeration—compression and absorption—were developed in the nineteenth century. All refrigeration is based on the thermodynamic properties of liquids and gases. It is not about adding "cold"—there is no such substance— but about pumping heat away. Refrigeration exploits the fact that when liquids turn into gases, heat is transferred away, like the steam rising from a bowl of soup as it cools down.

Since ancient times in Egypt, the principle of evaporation had been used to chill water: liquids were stored in porous earthenware jars, well wetted on the outside. As the surface water evaporated, it transferred heat away from the water inside the jar. In India, this technique was actually used to make ice. Trenches were dug and covered with straw. Shallow earthenware pans were put inside, filled with water. Under the right climatic conditions—calm and not too windy—the water turned to ice.

From the eighteenth century on, a series of inventors experimented with ways to speed up the chilling effects of evaporation. In the early nineteenth century, Richard Trevithick, a Cornish engineer, succeeded in building the first machines in which expanding air under pressure converted water to ice.

Air, however, was not a great refrigerant—it is a poor conductor of heat, which is after all the aim of the game. Engineers started to try different refrigerant gases. In 1862, the Harrison-Siebe vapor compression ice maker was launched, using ether instead of air. It was a vast and intimidating machine, "driven by a steam engine of fifteen horse powers." It worked on the same basic principle of the fridges in most of our kitchens. A gas—in this case, ether—is compressed through metal coils into a liquid state. It is then allowed to expand again into a gas, which causes heat to be removed—this is the refrigerant effect. Finally, the gas is re-liquefied, and the whole process starts again. The Harrison-Siebe machine worked very well, once early tendencies to explosions were resolved. The great steam-powered ice factories of the 1890s used the compression technique to churn out hundreds of tons of clean, diamond-bright ice a day.

This was not the only way to manufacture ice, however. French inventors, notably Ferdinand Carré, had come up with an alternative method: gas absorption. The difference is that instead of forcing the gas through compressor coils, it is dissolved in a "sympathetic" liquid. In Carré's version, the liquid is water and the refrigerant is ammonia. It is a more complex process than compression: instead of one substance to consider, there are two. Nevertheless, Carré's machine was impressive. It worked on a continuous cycle and in 1867 could produce as much as 200 kg of ice per hour. In America, in the Southern states, which had never had a dependable supply of natural ice, factories sprang up equipped with vast Carré absorption machines. By 1889, there were 165 plants in the South, manufacturing beautiful, clear, artificial ice, with which to chill mint juleps or aid the transport of fragile Georgia peaches.

Yet while the commercial ice industry had become mechanized, the ordinary American housewife still struggled on with her icebox. As late as 1921, a writer for *House Beautiful* was complaining of the drudgery of maintaining this cold receptacle:

> Somebody has had to wipe up the wet spot where the ice man set the cake while he was waiting . . . Somebody has had to pull out the pan each day from underneath and empty out the water . . . Somebody has had to keep smelling around the ice-box, day by day, to see when it began to get foul and needed scouring.

All of this daily tedium was dispensed with upon the arrival of domestic fridges, both gas and electric, which happened in the interwar years. It has been said that the decade between the end of World War I and the start of the Depression saw the most "dramatic changes" in the patterns of household work in America of any decade in history. In 1917, only 25 percent of households in the United States were on the electric grid. By 1930, that number was 80 percent. A critical mass of consumers with access to electricity was a crucial factor in the spread of the electric compressor refrigerator. This was a high-stakes business: unlike an electric iron or an electric kettle, an electric refrigerator is never switched off; for twenty-four hours a day, every day, it uses power, humming away. The electric companies therefore had a strong interest in encouraging the spread of electric refrigeration in the home.

The first household names in refrigerators were Kelvinator and Frigidaire, both firms founded in 1916 and both electric. There were teething troubles, to put it mildly. If you bought an electric refrigerator in the 1910s, you did not get an entire self-contained unit. The fridge company came and installed a refrigerating mechanism in your existing wooden icebox, which often could not take the strain, warping and yanking apart as the motor rattled away. The machinery, moreover, was so huge, it might leave scant room in the icebox for food. To get around this, the compressor and the motor were sometimes installed in the basement, with the refrigerant laboriously pumped back upstairs to the icebox. The compressors frequently malfunctioned, and motors broke. More worrying was the fact that the early refrigerant gases used—methyl chloride and sulfur dioxide—were potentially lethal. Given that the units were very badly insulated, this was a serious risk. In 1925, the scientist Albert Einstein decided to design a new, better refrigerator, after he read a newspaper story about an entire family killed by the poisonous gases leaking from the pump of their fridge. The Einstein refrigerator—developed with his former student Leo Szilard and patented in November 1930—was based on the principle of absorption, like the Carré machines. It had no moving parts and needed only a small heat source such as a gas burner to make it work.

The Einstein fridge was never marketed because it was overtaken by events. In 1930, the industry introduced a new nontoxic refrigerant called Freon-12. Almost immediately, all new domestic fridges adopted Freon. It seemed a new dawn—though half a century later, refrigerator manufacturers would be frantically looking for alternatives to Freon, because it is one of the main chlorofluorocarbons implicated in the depletion of the ozone layer.

Also in 1930, US sales of mechanized fridges overtook sales of iceboxes for the first time. By then, fridge design had moved far beyond those old leaky wooden chests. The early self-contained fridges of the 1920s tended to be white, with four legs like a dresser. The most famous was perhaps the General Electric "Monitor Top" refrigerator, a white box on legs with the refrigerating mechanism in a cylinder protruding on top. By the 1930s, fridges grew in stature and lost the legs, developing a streamlined metallic beauty.

In 1926, Electrolux-Servel devised a continuous absorption gas-powered fridge, and for a while it looked as if gas fridges might be preferred to electric ones. The basic invention came from two Swedish engineers, Carl Munters and Baltzar von Platen. These new gas fridges needed no motor to run, were cheaper, and ran silently. A Servel ad from the 1940s showed a well-dressed white couple boasting that they had managed to retain the services of their black maid after they bought an Electrolux: "Mandy's giving us another chance since we changed to silence." Mandy comments: "Lordy, it sure is *quiet!*" Despite the advantage of silence, Servel never had the same clout as the big electric companies like General Electric and now, the idea of a gas fridge sounds quaint. But the competition between the two models—silent gas versus humming electricity—drove innovation on both sides, which is partly why American fridges became so good, so quickly. Fridges of the late 1930s already had plenty of modern accoutrements: push-pull latches on the door; adjustable split shelving; hydrating compartments for vegetables and salads— all things that fridges today still sell themselves on.

What Frigidaire and Electrolux manufactured, America bought. In 1926, consumers bought 200,000 fridges (at an average price of $400); by 1935, sales had jumped to 1.5 million (averaging $170). Nearly half of the country now owned a mechanical fridge. Ads encouraged consumers to think of the fridge as a place from which wondrously fresh foods emerged. Kelvinator pushed the idea of "Kelvinated" foods:

> Taken from the cold frosty air of a Kelvinator-chilled refrigerator, they are irresistible. Think of sliced oranges, served ice-cold; of cantaloupe or grapefruit, chilled through and through; or of home-canned fruits, served cold in their rich juices. Think of the cream for your cereals cold and refreshing.

Older methods of food preservation did not, on the whole, purport to improve food, but merely to rescue it from harm. People knew that red herrings were not as good as fresh herrings; but you'd rather have red herrings than rotten herrings. By contrast, the fridge industry claimed not just to preserve food, but to transform it.

The reality was not always so appealing. One common complaint about fridges was that they made food taste off even when it was fresh. In 1966, a food storage expert, R. C. Hutchinson, noted that consumers believed that refrigerated foods "lose much of their flavour and acquire another taste." From a business point of view, however, this wasn't necessarily a problem; it was also an opportunity. Fridges gave rise to new storage products such as plastic wrap (invented in 1953 as Saran Wrap) and Tupperware (first sold in 1946). "Hear that whisper?" urged an ad of the 1950s. "That's Tupperware's airtight promise to keep food flavor fresh!"

Tupperware was also promoted as a storage device for frozen foods, an aid to cramming the maximum produce into the limited space of a home freezer. By the time that Tupperware launched, frozen food was heading toward being a billion-dollar industry, though it had a slow start. American fridges of the 1930s were hopeless when it came to freezing. Frozen items had to be stored in a tiny space next to the evaporator coils, where the fridge was coldest; there was only room for a package or two of food, and ice cubes had a nasty habit of melting and fusing into a single block.

The potential for frozen food improved dramatically with the introduction of the "two-temperature refrigerator" in 1939—the refrigerator-freezer. At last, cartons of ice cream and ice cubes could be kept entirely separate from the contents of the fridge and at a consistent subzero temperature. Another innovation was that the evaporator coils were now hidden within the refrigerator walls, which made for more even refrigeration but also got rid of the nightmare of defrosting. The household in possession of such a machine had every reason to fill it up with some of the frozen bounty now available: frozen orange juice concentrate so that the family could have "fresh" juice every morning (this was the single-most-successful commercially frozen product in postwar America, with 9 million gallons sold in 1948 to 1949); strawberries, cherries, and raspberries, so that the fruits of summer could be enjoyed even in midwinter; newfangled fish sticks; and frozen peas, courtesy of Birds Eye.

Clarence Birdseye, who created the modern frozen-foods industry in the 1920s, liked to say that there was "nothing very remarkable about what I have done . . . the Eskimos had [frozen foods] for centuries." This was too modest. It is true that ice had been used— and not just by the Eskimos—to preserve fish and meat long enough for it to be brought to market. But only Birdseye developed a quick-freezing technique so delicate it could be applied not just to carcasses of flesh but to tiny green peas.

Like the United States, Russia was a country of vast distances and icy winters, which encouraged the use of freezing to preserve food. In 1844, Thomas Masters, an ice expert from Britain (a small country with mild winters), wrote of the wonders of the ice market at St. Petersburg, "containing the bodies of many thousands of animals in a state of congelation, and piled in pyramidical heaps; cows, hogs, sheep, fowl, butter, fish—all stiffened into stony rigidity." The produce was all frozen solid. If you wished to buy anything, it would be chopped up for you "like wood."

Clearly, this is a very different proposition from a bag of frozen petits pois, pre-podded and on the table in five minutes. The St. Petersburg ice market was selling rugged, survivalist food, a far cry from the American housewife of a hundred years later, who took a Swanson TV dinner from the freezer and, scarcely pausing in the transition from cold to hot, transferred it to her electric oven. Clarence Birdseye's innovation was to create frozen foods that could fit seamlessly and hygienically into twentieth-century suburban living; no ice pick required.

Birdseye was a fur trapper who had previously worked as a biologist for the US Department of Agriculture. His invention came out of a simple observation. He and his wife, Eleanor, and their baby son, Kellogg, were on a fur-trapping mission in Labrador, in northeast Canada, on and off during 1912 to 1915. They lived in a tiny shack far from the nearest settlement. For food, they survived on fish and game, frozen in the Arctic winds. Birdseye noticed that their food—rabbit, ducks, caribou, fish—tasted better in winter than in spring and autumn. The fast-frozen winter meat tasted as good as fresh. The reason, he assumed, was that it had frozen more quickly. He also experimented with freezing green vegetables, which were only shipped very infrequently to Labrador. Birdseye found he could quickly freeze cabbages and other greens by plunging them in barrels of salt water. He even went so far as to use little Kellogg's baby bath to aid the process.

Traditional methods of freezing food—such as those of the St. Petersburg market—involved simply burying the food in ice or snow, where it froze slowly. This encouraged large ice crystals to form and badly affected the quality of the food, damaging its cellular structure. When slow-frozen food thawed, fluids leaked out. The problem was especially pervasive with meat. In 1926, the *London Times* complained of the "copious" quantities of "bleeding or drip" that tended to exude from slow-frozen beef as it thawed.

The solution was at hand. When Birdseye returned to the United States from Labrador in 1917, he had an initial investment of just $7 for an electric fan, some cakes of ice, buckets of brine, and fillets of haddock. He set to work in a corner of a New Jersey ice-cream plant,

trying to "reproduce the Labrador winters" in America. By 1925, Birdseye had developed an entirely new method for quick-freezing food,
using metal plates chilled in calcium chloride solution to −45°F.
Packets of food were pressed between the metal belts and froze almost instantly—far more quickly than in any previous technique. At
first, Birdseye used the method to freeze fish, establishing the General Seafood Corporation in 1925, the idea being that it would become the General Motors or General Electric of frozen food. In
1929, he sold his company and patents for $22 million to Goldman
Sachs and the Postum Company.

The frozen-foods business was not an instant success. Early
frozen peas did not taste good. Only in 1930 was it found that peas
and other vegetables needed to be blanched in hot water before they
were frozen, to inactivate the enzymes that made them spoil. The
unreliable quality of frozen foods contributed to the deep suspicion
with which many shoppers viewed them. There was a general sense
that frozen food was subpar: salvaged goods. The turning point was
when Birdseye embarked on a PR campaign, renaming the produce
as "frosted foods," a name that implied icy glamour. "Frozen food"
was something you would eat rather than starve. "Frosted food" was
the stuff of childhood fantasy. It worked. As of 1955, the frozen-food
market was worth $1.5 billion in the United States.

Frozen foods became popular in Britain, too. Green garden peas
would surely have never become such a central part of the British
diet without the deep freeze. Sausage, chips, and peas; chicken,
chips, and peas; pie, chips, and peas: most of the vegetable fare on a
pub menu arrives courtesy of Birdseye. In 1959, sales of frozen peas
overtook sales of fresh peas in the pod in Britain for the first time.
The strange thing was that British shoppers eagerly purchased
frozen foods, despite the fact that they had nowhere to store them.
The *London Times* noted that this was a "handicap" for housewives
who "need suddenly to enlarge a meal for extra guests when shops
are closed." Frozen-food manufacturers looked into creating after-

hours frozen-food vending machines, to answer this predicament, but so far as I am aware, none were ever operational. Picture the scene: hundreds of frantic housewives queue around the block to pick up emergency supplies of frozen Chicken Kiev to deal with the sudden arrival for dinner of their husband's boss. As late as 1970, the number of households with access to a freezer of any kind stood at just 3.5 percent. For the rest of the country, any frozen food purchased had to be crammed in a tiny space on top of the ice-cube tray. I remember it well: a half-eaten cardboard container of raspberry ripple ice cream, forming large ice crystals as it melded itself onto the ceiling of the fridge.

A gulf had opened between refrigerated America and the rest of the world. It was a question of culture, as much as of the capital outlay required to purchase fridges and freezers. For a long time, Europeans actively rejected the technology of cold storage.

The French had a word for it: *"frigoriphobie,"* meaning fear of fridges. Both consumers and producers at Les Halles, the main food market in Paris, held out against refrigeration. Buyers feared it would give tradesmen too much power over them: fridges would enable them to take old food and pass it off as fresh. The sellers should have welcomed the technology—after all, refrigeration gives traders a longer window in which to sell their food. Yet when the fridge was introduced to Les Halles, traders reacted as if they had been personally slighted. The fridge, they insisted, was like a mausoleum in which the true nature of a great cheese would be killed. And who is to say they were entirely wrong? A fridge-cold Brie is a dull thing compared to the oozing wonder of a Brie matured to its peak in an old-fashioned larder.

Nor were Continental consumers eager to use refrigeration in the home. Given their patterns of food shopping, there was frankly no need. In the 1890s, American icebox manufacturers made some tentative steps into the European market, asking American consuls for information on local demand for iceboxes. The intelligence they received was not encouraging. The consuls replied that in the great cities of southern France, meat was butchered every day in winter, twice a day in summer. The majority of people went shopping for food twice a day, eating up everything they bought. For as long as women were happy to shop and cook as often as this, and for as long as sellers could provide them with the fresh produce they needed, iceboxes were surplus to requirements.

In Britain, too, there was no rush to buy fridges. For much of the twentieth century, American visitors to Britain found that everything was the wrong temperature: cold, drafty rooms; warm beer and milk; rancid butter and sweating cheese. In 1923, an article in *House and Garden* noted that "refrigerators, which are a commonplace in American households, are not sufficiently known or used over here." Given the unreliable and poisonous nature of 1920s fridges, being slow to adopt them may have been no bad thing. Yet the British antipathy to fridges was not entirely rational. Long after electric refrigerators had become safe and consistent, and long after the majority of houses were electrified, there was still a view that they were wasteful and decadent. Frigidaire noted the challenge of cracking the British market: "The hard sell was probably essential in a Britain which regarded ice as only an inconvenience of winter-time and cold drinks as an American mistake." This fear of the excess of American appetites was a national austerity of the mind that long predated the actual austerity of wartime and its aftermath. As of 1948, just 2 percent of British households owned a fridge.

Eventually, the British got over their aversion to cold. Fast forward to the 1990s, and the average British household owned 1.4 "cold appliances" (whether a refrigerator, a refrigerator-freezer, or a chest freezer in the garage). There was a seemingly insatiable desire for Smeg "Fabs," retro refrigerator-freezers in pastel colors with big clunky handles, like American fridges of the 1950s. In other words, by the late 1990s, the British had just about caught up with where Americans were fridgewise in 1959.

The design of a fridge reflects the kind of lives designers think we lead and the kind of people they think we are. In 1940, an American refrigerator salesman commented that "fifty per cent of our business is preserving women, not fruit." Something like a push-pull door handle with three-way action was important because "it makes a lot of difference to the woman whether she can walk with her arms full of something." Fridges sold themselves to women based on desire—dreamy pastel colors—but also on duty: customers were told that it was their job to keep the family's food safely cold.

By the mid-1930s, more fridge compartments were added—removable split shelving, vegetable hydrators—encouraging households to keep a higher percentage of their food refrigerated. In all of this, the original purpose of cold storage, to keep food in optimum condition for longer, was often lost. Fridges came, and still do, with tidy egg containers. Yet these indented trays do a worse job of protecting eggs than the cartons they are sold in, which shield eggs from picking up other odors. Moreover, in a cool climate, eggs are better stored out of the fridge, at least if you use them up quickly. A room-temperature yolk is less likely to break when you fry it; and not so liable to make cake batter curdle. But your room temperature may not be the same as mine. In America, unrefrigerated eggs are viewed as hazardous objects; and so they are, in the hotter states during the warmest months. A 2007 study from Japan found that when salmonella-infected eggs were stored at 50°F over six weeks, there was no growth in the bacteria. Even at 68°F, there was negligible bacterial growth. At temperatures of 77°F and above, however, salmonella growth was rampant. In Alabama, in July, an unrefrigerated egg could be lethal.

The internal dimensions of the fridge continue to evolve. In the 1990s, the internal shelving in British refrigerator-freezers tended to be boxy and geometrical, reflecting the fact that large sectors of the population were living off neat rectangular boxes of precooked meals. In recent years, an appliance expert told me, this has changed. People want multiple vegetable and salad crispers and more varied shelving, a reflection of the fact that they are returning to "scratch cooking" (which is "cooking" to you or me). Internal wine racks have also become common.

Refrigerators started as devices for helping us to feed ourselves safely. Yet they have become insatiable boxes, which themselves demand to be fed. Many foods now considered staples principally came into being in order to give people something to put in their new fridges. I don't just mean the obvious things, like fish fingers and frozen French fries. Take yogurt. Before World War II, yogurt was hardly eaten in the West. Although a traditional food in India and the Middle East, where it was made fresh as needed and kept in a coolish place, fermenting and clotting over time, yogurt had zero commercial potential in Britain or the United States. Without fridges, people consumed their dairy desserts mostly in the form of homemade milk puddings, made fresh and served warm: rice pudding, sago, tapioca (which British schoolchildren referred to as frog spawn, on account of its texture). From the 1950s onward, consumption of milk puddings fell dramatically year after year. Meanwhile, yogurt was growing into a multi-billion-dollar global industry. Why? You could say tastes had changed, but this still doesn't explain why warm rice pudding with a dollop of strawberry jam should suddenly be spurned and why cold strawberry yogurts in plastic containers should suddenly be embraced.

So much of what we think of as personal taste is actually a consequence of technological change. Yogurt manufacturers were capitalizing on the fact that, having bought a shiny new fridge, the owners wanted plenty of things to put in them. Those neat little pots looked good lined up on the split shelving; how they tasted was almost irrelevant (some yogurts were nice enough, but many were blander and more sugary than the traditional puddings they replaced). For the first time in history, almost everyone had access to ice all year round. Sometimes we just didn't know what to do with it.

~ *Molds* ~

MRS. MARSHALL SOLD ICE-CREAM MOLDS shaped like apples, pears, peaches, pineapples, bunches of grapes, towers of cherries, giant strawberries, ducks, hens, swans, and fish, as well as the more abstract bombes, domes, and pillars. Her molds came in affordable pewter or tin, or best-quality copper for jellies.

Molding something is a way of imposing your will on ingredients with great force. The shapes of food molds are culinary technology at its most capricious. By what logic did the Indian ice cream, *kulfi,* that dense confection of cooked milk, come to be made in cone-shaped molds? Why not square? Or hexagonal? No one seems to know. The answer is always: "It is traditional."

Some food molds follow a certain logic: fish mousse goes into fish-shaped molds and melon-flavored ice might be packed into a cantaloupe mold. Often, however, there is no sense behind the shape, except for the taste and mores of the times. The Turk's head was a popular patisserie mold in the early twentieth century, mimicking a turban; it's a pretty shape, but the idea behind it—eating a Turkish man's head—seems in very poor taste now.

Molds are driven by fantasy and a desire for the spectacular, and our sense of spectacle changes over time. Medieval gingerbread molds, hand carved from wood, might depict harts and does, wild boars and saints. The stock of images available to us now is far larger; but our imaginations are often smaller. In kitchen shops today, you can buy a large cake mold resembling a giant cupcake.

8

KITCHEN

I think it is a sad reflection on our civilization
that while we can and do measure the
temperature in the atmosphere of Venus we
don't know what goes on inside our soufflés.
NICHOLAS KURTI, 1968

IN THE EARLY YEARS OF THE NEW MILLENNIUM, DESIGNERS liked to joke that a house was just a kitchen with a few rooms attached. In 2007, before the Great Recession, the *New York Times* identified a new cultural malaise: professionals suffering from "post-renovation depression" when their kitchen project is finally done and they have to stop obsessing over the minutiae of faucets and backsplashes. "People said it would be a great relief when it was over," commented one homeowner whose elaborate kitchen refit nearly doubled the size of her house. When the room was finished, she complained, it "left a huge hole in my life." Such unhappiness might seem strange to the Victorian housemaid embarking on the wearying daily chore of cleaning and blacking a cast-iron stove. The expensive kitchens of the present speak of a degree

of comfort, particularly for women, that is historically unprece-
dented. The technology of the kitchen is both cause and consequence of that comfort. Our lives are comfortable because we have
upscale fridges and toasters. We buy the fridges and the toasters to
fit with our comfortable lives.

The luxury and sheen of the modern showroom kitchen would
have been a foreign land to our forebears a hundred years ago, when
electric refrigerators were unknown and the gas stove an exciting novelty. How futuristic these rooms would surely seem to them: the
panoply of "clever storage" devices, the hissing espresso machine, the
cavernous refrigerator-freezer, the color-coordinated cabinets and
mixers. How could you explain to an Edwardian newlywed just getting to grips with her mahogany cabinet refrigerator and her case of
silver-plated knives that the day would come when people—men as
well as women—would treat kitchen remodeling as a hobby, when
perfectly good electric blenders would be thrown away because they
didn't quite match the slate blue of a new set of kitchen cabinets?
How has it become normal that on moving to a new house, you would
rip out the entire kitchen put in by the previous owners—perhaps
only a few years earlier—and install your own from scratch, with all
new fixtures and fittings: new range, new floor, new kitchen sink?

If you look beyond the granite and the glass and the recessed
LED lighting, however, there is surprising continuity between the
technology of today's kitchens and those of the past. In the 1890s,
French chemist Marcelin Berthelot predicted that by the year 2000,
cooking would be finished and human beings would subsist on pills.
This meal-in-a-pill idea has been a perennial feature of our space-
age fantasies. Yet despite all the encroachments of industrial food—
despite Slim-Fast and "breakfast-in-a-bar"—the business of cooking
persists. Even the food eaten on the early space missions did not
generally come in pill form. The further they got from Earth, the
more it seemed astronauts craved the tastes and textures of home.
The meals may have been freeze-dried, but they were approxima-
tions of the stews and puddings of normal kitchens. According to

Jane Levi, a historian of space food, one of the key discoveries of Project Gemini—the ten manned space flights that NASA conducted between 1965 and 1966—was that astronauts don't like cold potatoes.

However radical we may think we are in our everyday beliefs, when we step into a kitchen, most of us become (small "c") conservatives. We chop food with knives, stir it with spoons, and cook it in pots. As we stand in our modern kitchens, we still use the colanders, the pestles, and the frying pans of the ancients. We do not start from first principles every time we want to produce a meal but draw on the tools and ingredients we have at hand, governed by the rules and taboos and memories we all carry in our heads about cuisine.

Some people don't like this. For French scientist Hervé This, one of the inventors of the term "molecular gastronomy," our cooking is guilty of "technical stagnation." In 2009, This asked:

> Why do we still cook as we did in the Middle Ages, with whisks, fire, and saucepans? Why this outdated behavior; when, at the same time, humanity is sending probes to the outer limits of the solar system?

So why are we so resistant to change in the way we cook? One reason is that experimenting with new foods has always been a dangerous business. In the wild, trying out some tempting new berries might lead to death. A lingering sense of this danger may make us risk-averse in the kitchen. But our attachment to certain ways of cooking goes beyond self-preservation. Many tools have endured because they work so well. Nothing does the job of a wooden spoon better than a wooden spoon. There's also the fact that when we pick up a certain utensil to cook a certain dish in the traditional way—

whether it's a Valencian paella made in its wide shallow pan or a Victoria sponge made in old-fashioned sandwich cake pans—we are enacting a ritual that binds us to the place we live and to those in our family, both living and dead. Such things are not easily shrugged off. As we have seen, every time a new cooking technology has been introduced, however useful—from pottery to the microwave to the smokeless stoves of the developing world—it has been greeted in some quarters with hostility and protestations that the old ways were better and safer (and sometimes, in some ways, they were).

Hervé This says that there are two types of technological change: local and global. The small local changes in kitchen machinery are the easiest to accept. The example This gives is an improvement to a balloon whisk that adds more wires to whip eggs more efficiently. New gadgets feel safest when they remind us of other objects that we already know. This explains why early refrigerators looked like heavy wooden Victorian kitchen cabinets and why lemon squeezers of the 1860s were often clamped on the table like hulking iron meat grinders. In the 1950s, countless utensils took the form of a continental Mouli food mill with a rotating handle: suddenly, there were rotary cheese graters and rotary herb mills and they were greeted with enthusiasm. Although—unlike the Mouli itself—neither is really a very good tool on its own terms: the herbs get mashed and the cheese grater always leaves a thin slice behind in the rotary drum. But at this time, a rotating mechanism felt natural and this was what mattered. Hands and brains were accustomed to the action of processing food through a drum with a circular motion.

It is far harder to accept a technology that is entirely new. This is what This calls "global" change: the sort of shift that came about when our ancestors decided to start cooking things in pottery or when

Count Rumford rejected the idea that an open hearth was a good way to heat food. Such changes disturb our natural conservatism.

Take egg whites. Instead of just tinkering around the edges by adding more wires to a preexisting whisk, a global shift in technology would call into question why a whisk is used for beating egg whites at all. This is what Dr. This wants to know. "Why not use, instead, a compressor and a nozzle that can introduce air bubbles into the egg white?" Or why not come up with some entirely new device that no one has thought of yet? Why not use all your ingenuity and imagination?

Well, why not? For most people, cooking is a laborious enough process without adding the creation of new tools to the mix. The past couple of years have seen a small renaissance in home cooking, partly a response to the austerity of the recession. But if you look at the past forty years, the picture is one of a radical decline in cooking. When British chef Jamie Oliver went to northern England, to the town of Rotherham in 2008 for his TV program *Ministry of Food,* he met people who owned electric ovens but had no idea how to switch them on. A 2006 survey by the Institute of Food Technologists found that although 75 percent of Americans ate dinner at home, fewer than one-third were cooking their meals from scratch. The real culinary breakthrough would be to get the remaining two-thirds of the population to cook with whisks, fire, and saucepans, rather than persuading them that these technologies are outmoded. The act of whipping egg whites with a whisk may seem unadventurous, but the cook holding the whisk has had to overcome numerous obstacles to continue being one of those who engages with cooking technology at any level. The noncooking majority don't get anywhere near this far. To repeat Dr. This's question, why don't cooks reinvent the whisk? There are a hundred reasons why not, ranging from "It's not how my mother did things" to "I don't have all the time and resources in the world" to "My balloon whisk works fine."

In recent decades, however, there has been a movement in cookery that endlessly asks "why not?" Why not serve ice cream hot instead of cold? Why not vacuum seal eggs in a plastic bag and "scramble" them in a sous-vide bath? Why not take mayonnaise and fry it? This movement has gone under many names: molecular gastronomy, techno-emotional cookery, hypercuisine, vanguard cooking, modernist cuisine. Whatever you call it—and I will stick with "modernist" for now—this movement represents the biggest rethink of kitchen technology since the microwave (a gadget that the modernists adore).

When Nathan Myhrvold wants to eat a hamburger, he doesn't take down a trusty cookbook or try to remember some advice his mother once gave him. Nor does he mindlessly sling a burger patty on a griddle. He first works out exactly what he wants from a burger—the "ultimate" hamburger. Myhrvold likes his meat to have a rosy-red inside with a dark, caramelized exterior. This wouldn't be everyone's "ultimate" hamburger, but it is his. And it turns out to be nearly impossible to achieve with conventional cooking methods. On a griddle, by the time the outside is brown enough for Myhrvold's liking, the inside tends to be gray and overcooked. So Myhrvold used some of his immense wealth (in a previous life he was chief technology officer for Microsoft) to experiment until he came up with a technology that gave him the results he desired.

The answer, it transpired, was far from obvious, not to say impossible to achieve in any normal domestic kitchen. To stop the inside of the burger from overcooking, you have to first dunk it in liquid nitrogen to cool it down fast. To ensure the caramelized exterior, you then deep-fry it in hot oil for exactly one minute: long enough to brown all the surfaces, not so long that the heat penetrates the center. But before either the liquid nitrogen or the hot oil, Myhrvold applies yet another technology, cooking his burger long and slow in a sous-vide machine: for about half an hour in tepid water—to ensure perfect medium-rare tenderness.

Sous-vide is to modernist cooking what spit-roasting was to the Elizabethans: the default technology for cooking almost everything. The name comes from the French for "under vacuum," and the process means cooking vacuum-sealed food at precisely controlled temperatures in water. Food is vacuum-packed in heavy-duty plastic bags, then suspended in a low-temperature water bath, sometimes for many hours (cheaper cuts of meat may take as long as forty-eight hours to tenderize). In principle, sous-vide is a bit like the slow cookers that have been around for decades, or the bain-maries of which the Victorians were so fond, but the overall effect is entirely novel. To someone reared on homespun meals, sous-vide hardly seems to be cooking at all. The food in its plastic looks alarmingly like medical samples, or brains in formaldehyde. Another unsettling aspect is the total absence of aroma. Sous-vide enthusiasts boast that the food's scent is all locked up in its silent bag. With sous-vide, you have none of the normal sensory cues that a meal is cooking: the smell of garlic sizzling in oil, the *blip-blip* of risotto in a pot.

I was a sous-vide skeptic. I didn't like the aesthetics, the wasted plastic and lack of romance. To cook sous-vide, you need two separate pieces of apparatus: two more possessions to add to our already overequipped kitchens. First, a vacuum-sealing machine, which looks more like a laser printer than anything to do with food. It is a plastic rectangle with buttons on the top. Ingredients are placed in thick plastic pouches. The edge of the pouch is fed into the vacuum sealer, which sucks most of the air out, leaving the food shrinkwrapped. The second component of sous-vide is the water bath, a stainless steel vat. It gets filled with water and set to very precise temperatures using a digital control panel, ready to cook the shrinkwrapped food.

I didn't want this giant metal *thing* on my countertop. Then I borrowed one from the main seller of sous-vide for the UK domestic market (Sous Vide Supreme) and discovered that food cooked by sous-vide was qualitatively different from food cooked by any other technology. Not that it was always better. Misjudging the cooking temperatures and times with sous-vide is disastrous because you can't check the progress of the food as it cooks in the same way that you can using a pan. You set the water bath to the required temperature, vacuum seal the food in the bag, submerge the bags, set the timer, and wait for the bleep. No stirring, no basting, no prodding or tasting. No human input at all.

When you get it right, however, sous-vide food is extraordinary, even hyperreal. Fruits and vegetables that might otherwise be boiled, steamed, or poached assume a concentrated flavor. Globe artichokes tasted, almost overwhelmingly, of artichoke; I could feel that odd soapy taste on my tongue almost an hour after eating them, because none of the flavor compounds had leached out into the water, as they do in a pan. Sous-vide apples and quinces, cooked for two hours at precisely 181.4°F, were deeply fragrant and golden, with a far better texture than I've ever achieved by poaching: dense but not too grainy, like the essence of autumn. Carrots with rosemary seemed to be infused through every cell with the herb. And potatoes! For years, I have fantasized about some perfect boiled potatoes that I once ate as a child on holiday in France: firm, yellow, and buttery, the platonic ideal of potato. Little did I think that they would one day emerge from a plastic bag in my own kitchen.

The sous-vide machines aimed at domestic kitchens sell themselves mostly as devices for cooking meat. The images on the box are all of fillets, chops, and steaks. A kitchenware buyer told me that this was because "meat and steak are such an investment," and most people (vegetarians aside) do not want to lavish so much money— $400 and up—on a tool for cooking vegetables. It is also true that sous-vide meat and fish yield results that are unique and new. You can take tough cuts of meat and, by cooking them at the lowest possible temperature—just high enough to denature the proteins and kill off pathogens, but no higher—achieve a tenderness that would once have been unimaginable. For the first time, you can cook proteins with minimal loss of juices. Chewy flank steak now melts like mousse. Cuts that were tender anyway, like pork fillet, become alarmingly soft, almost jellylike. A traditional pan-fried steak cooks in gradations as the heat gradually conducts through the meat, from a very cooked exterior to—if you are lucky—a pink inside. Sous-vide proteins are different: they are cooked to the same degree all the way through. And in contrast to the meat cookery of the past, you sear it after it has cooked, not before (without this final searing stage, sous-vide meat is pallid and moist).

Sous-vide technology was first invented for industrial food in the 1960s by French and American engineers working for Cryovac, a food packaging business. It was initially seen as a way to prolong shelf life, and indeed vacuum sealing is still widely used for that purpose in the food business. It was only in 1974 that a chef first found that Cryovacking could make food *better* rather than just longer lasting, if the vacuum-sealing technology could be combined with slow, gentle cooking. In France, the triple-Michelin-starred Pierre Troisgros was dissatisfied with his methods of cooking foie gras—swollen goose or duck liver at that time being considered an essential component of any Michelin-starred restaurant. Troisgros found that the foie gras was losing up to 50 percent of its original weight when it was sautéed. Troisgros consulted an offshoot of Cryovac, a school called Culinary Innovation, and was advised to shrink-wrap his foie gras in several layers of plastic before slow-cooking it. It worked. The loss of weight was reduced to 5 percent, thus saving Troisgros a fortune. The liver tasted better, too (at least to those who like foie gras). The fat that had previously melted away in the pan was now retained, giving the liver exceptional richness.

Six years earlier in Britain, a Hungarian physicist named Nicholas Kurti had been making some discoveries of his own. In 1968, Kurti gave a Friday-night lecture at the Royal Institution, entitled "The Physicist in the Kitchen." Kurti found it very sad that the role of science in the kitchen had not been given more attention. He showed the audience a series of hypodermic syringes, which he used with a dramatic flourish to inject pineapple juice into a loin of pork to tenderize it (pineapple contains an enzyme, bromelin, that breaks down proteins). He used a microwave oven to construct an "inverted Baked Alaska" with an outer layer of chocolate ice cream encasing an inner layer of hot meringue and apricot puree. Finally, Kurti brought in a leg of lamb cooked for eight hours at precisely 176°F until fantastically tender: here already was a version of the sous-vide idea of cooking meat low and slow at highly controlled temperatures. Kurti is now revered among the modernist chefs and food scientists as one of the fathers of high-tech cuisine.

In the 1960s and 1970s, however, the food culture was not quite ready for hypodermic syringes and vacuum packing. Sous-vide was widely practiced in the catering industry, but it was a dirty little secret. Many of us have eaten sous-vide without knowing that was what it was. If a caterer needs to make something like coq au vin for a corporate dinner for 200, sous-vide is highly convenient, because the dish can be portioned up in its bags, then entirely precooked in the water bath, and reheated as required. It cuts down on labor costs, too. But it was not something chefs generally boasted about. As recently as 2009, there was a scandal when British chef Gordon Ramsay was "exposed" for serving "boil-in-the-bag" food at some of his restaurants.

Sous-vide has really only come out in the open over the past couple of years as part of the rise of modernist cooking. Now, restaurants will advertise the fact that they have used sous-vide to compress watermelon, "flash pickle" celery, or reinvent hollandaise. The shame surrounding this technology has been replaced with pride. It has gone from being an indication of thoughtlessness to a sign that culinary experts have gone to great trouble to make an ingredient taste more intensely of itself. Sous-vide is just one of the bewildering tools in the modernist kitchen, along with cream whippers installed with liquid nitrogen canisters for making foams and "espumas," and crazily powerful homogenizers for making "nano-emulsions." Around the world, chefs are wielding freeze-dryers and centrifuges, Pacojets and siphons. And, like children at play, they are constantly asking, "why not?" Instead of cooking something on a hot griddle, why not put it on an Anti-Griddle, whose surface chills food to −22°F, so cold that the surface takes on a crispy texture, as if it had been fried?

In the professional kitchens that have adopted them, these high-tech tools have wrought profound changes in the way chefs cook. In the old French gastronomy of Escoffier, chefs had a lexicon of techniques to fall back on, etched indelibly on their memories. They knew when to use a sauté pan and when a casserole dish. By contrast, the new chefs keep questioning the fundamentals of culinary technology. At El Bulli, Ferran Adrià took nothing for granted about how food should be prepared, shutting his restaurant for six months a year so that he could conduct rigorous experiments on the best way to cut salsify, for example, or how to freeze-dry pistachio nuts.

It remains to be seen to what extent modernist cooking techniques, significant as they are, either can or should translate to the domestic kitchen. There's certainly a place for sous-vide, but I can't see Anti-Griddles and centrifuges arriving in many homes. It would be exhausting to live like this, always questioning everything. Even the modernists cannot keep it up all the time; there are limits to how much you can deconstruct. Before the workday started at El Bulli in the morning, all of the chefs drank a cup—not of spherified melon or snail air—but of coffee. Liquid, not solid. Hot, not cold. Just like in any other kitchen, though it was probably better coffee. The happiest memories of many of the unpaid trainees who worked with Adrià were of the "family" lunches at which they ate such normal fare as spaghetti with tomato sauce or cauliflower with béchamel. Unlike art, food is not so easy to tear apart and reinvent. Modernist cooking can entertain, but can it nourish as a home-cooked meal can?

Perhaps this explains the strikingly critical attitude modernist cooks sometimes adopt toward mothers and their cooking. Mothers are mentioned nine times in the first volume of Myhrvold's *Modernist Cuisine,* never with praise. On the one occasion I met him, Myhrvold spoke warmly of his own mother and how she let him loose in her kitchen, aged nine, to cook his first Thanksgiving dinner, aided by an exciting volume entitled *The Pyromaniac's Cookbook.* Yet in his book, mothers are repeatedly criticized for holding "common-sense" notions about food that turn out to be wrong (such as cooking pork until it is well done). *Modernist Cuisine* does not tell us about any of the times when the culinary common sense of mothers has been proved right. In contrast to "culinary professionals," Myhrvold notes, mothers and grandmothers in the past were "only cooking for themselves and their families." Only! As if feeding those close to you were an act of no importance.

The modernist movement in cooking does not represent the only good way to make a meal. Even Nathan Myhrvold admits that some of the most delicious food now being served anywhere in America comes from the motherly kitchen of Alice Waters, the chef proprietor of Chez Panisse in Berkeley and great guru of the organic movement, whose cooking is of the old pots-and-pans variety. Waters does not own a microwave, never mind a sous-vide machine. Her approach to food starts not by asking "why not?" but by asking "what is fresh and good just now?" Waters does not feel the need to reinvent such things as corn on the cob, happily shucking the plumpest summer cobs and boiling for a few minutes in unsalted water. In 2011, Waters was asked on a radio program what she thought of the new wave of high-tech cooking. She replied that it didn't "feel real" to her. "I think there are good scientists and crazy old scientists that can be very amusing but it's more like a museum to me. It's not a way of eating that we need."

The disagreement between Waters and the modernists shows how different strategies for cooking can coexist in the kitchen now. In the distant past, the arrival of a new technology often wiped out an old one. Pottery supplanted pit ovens (except among the Polynesians). The refrigerator replaced the icebox. The case with the new modernist gadgets is different. Sous-vide will not kill off the griddle or the stockpot. We have countless options now available to us, both low-tech and high-tech. Do we wish to cook like a grandmother or like a mad scientist? Either way is possible. We might decide to splurge on a sous-vide machine. Or not. We might think we'd rather have delicious cooking smells than the world's juiciest steak. As Alice Waters says, there is no *need* to cook like the scientists. There are plenty of other ways to make a delicious meal in the modern kitchen. The thing that defines our culinary life now is not this or that technique, but the fact that we can choose from so many different technologies when we amble into our kitchens and ponder what to cook.

When considering the high-tech kitchen, it's easy to fixate on gadgets and to forget that the single biggest piece of technology on display in the modern kitchen is the kitchen itself, as a room. Many of our individual kitchen tools are ancient. In Pompeii, there were already familiar pots and pans, funnels, sieves, knives, pestles, and spoons. But there was nothing quite like our kitchens.

The majority of households, for most of history, have not possessed a separate, purpose-built room in which to cook. For the ancient Greeks, cooking went on in a variety of different rooms, as portable baking ovens and portable terracotta braziers were shunted from room to room. There was thus no kitchen in an architectural sense. A remarkable range of Greek kitchen utensils have been unearthed by archaeologists: casseroles and saucepans, cleavers, ladles, and even a cheese grater. But these impressive tools had no kitchen to house them. Most excavations have unearthed not the slightest trace of a fixed hearth or kitchen in Greek dwellings before the fourth century BC.

The Anglo-Saxons, too, often lacked kitchens, because many did their cooking outdoors, especially in the summer months. The kitchen ceiling was the sky; the ground was the kitchen floor. Odors and smoke dissipated in the open air. This was a freer, more open-ended way of cooking food than we have with our boxed-off kitchens, though it must have had considerable drawbacks in rainy weather, or when there was ice, wind, or snow. During the winter, these kitchenless households would have relied a good deal on bread and cheese.

In the cottages of medieval Europe, on the other hand, there were usually fixed indoor hearths, but the room that housed the cooking fire was living room, bedroom, and bathroom as well as kitchen. In a one-room dwelling, cooking was just another activity to fit in amid the dirt and crowding. The pottage in the cauldron on the fire was part of the furnishing of the room. This continued to be the standard way for the poor to live for centuries; for millions, it still is. Several of the seventeenth-century paintings and etchings of Adriaen van Ostade depict the lives of Dutch peasants. Grimy families are shown clustered around a hearth. Dogs yap in the background. Babies are suckled. Pots and pans and baskets of clothes lie strewn on the floor. Men smoke pipes. A cleaver hangs on the wall. It looks nothing like any kitchen we would know. But here and there are hints of cooking activity; bowls with spoons in them; a coffee pot; something warming in a pan. It need hardly be said that the food produced in such a room cannot have borne much relation to the ambitious dinner-party constructions of the modern cook. Nor can it have been easy to perform simple tasks we take for granted: to chop an onion or beat an egg.

Most peoples' cooking lives were entirely untouched by all the great innovations in kitchen technology of the eighteenth and nineteenth century: the clockwork jack, the automatic knife cleaner, the Dover eggbeater; all these passed them by. Why would you want an eggbeater if you never beat eggs? Except for the enclosure of the fire in a grate, not much changed in the culinary possibilities for the poor from ancient to modern times. Well into the twentieth century, poor Scottish and Irish cottagers still cooked their meals on a frying pan balanced over a grate, alongside wet boots and drying laundry. Tenement living in the towns could be even worse. Charlie Chaplin grew up in a derelict attic room shared with his mother and brother. The entire "stifling" room measured twelve feet square. In one corner was an old iron bed, shared by the three of them. Dirty plates and teacups crowded on a table. Chaplin recalled the stench, the "air . . . foul with stale slop and old clothes." The only means of cooking was a "small fire-grate" between the bed and the window.

In such one-room dwellings, the kitchen was both nowhere and everywhere. Nowhere, because the inhabitants lacked most of the things we would consider necessary for cooking: a sink for doing dishes, work surfaces, and storage. Everywhere, because there was no escape from the stink and the heat of the fire. Cooking is my favorite activity, but under such circumstances, I'd rather not cook at all. The phenomenon of people who live off nothing but takeout is not a new one. Beginning in medieval times, pie sellers were always a feature of British towns, serving those living in cramped one-up, one-down cottages where there was no kitchen as such.

Part of the luxury of having a kitchen is the ability to distance yourself physically from cooking when you choose. In rich European medieval houses, this was taken to an extreme by building wooden kitchens that were entirely detached from the main house. All of the food requirements of the household—the baking and cheese making as well as the roasting—could be carried out in this single, specialized building. Those living in the main house could enjoy the benefits of food cooked in a large kitchen without having to endure any of the fumes or grease, or the fear that the kitchen would catch fire and burn the house down. When these kitchens did burn down, as happened periodically, new ones could be built without disturbing the main structure of the house. The one great drawback of such outhouse kitchens was that food cooled down as it was ferried through into the dining room.

In other great medieval establishments, there were vast high-ceilinged stone-floored kitchens built as part of the main house. The biggest practical difference between these kitchens and the ones we have today was that they were communal, like the famous Abbot's Kitchen at Glastonbury, an octagonal room with a hearth vast enough to roast an entire ox. This was a space whose equipment needed to be able to answer the appetites of an entire community of monks. Our own built-in kitchens, aimed at feeding a single family, or in many cases a single person, look individualist by comparison.

But one room was seldom enough to contain the multiple culinary activities of the grand residences of centuries gone by. As of the 1860s, a British country house typically contained numerous rooms, each devoted to different aspects of kitchen work. It was like having an entire block of food shops under one roof. There was a pantry, or dry larder, for storing bread, butter, milk, and cooked meat: this room needed to be kept cool and dry; architects had to ensure that no hearths were built in adjacent walls. The wet larder was where raw meat and fish were kept, along with fruits and vegetables. In larger houses, there was a further game larder, with hooks for hanging the game and a marble dresser for preparing it. Other culinary rooms included the dairy for churning butter, making cream and cheese; the bakehouse, containing a brick oven to supply the household with bread; a smokehouse; and perhaps a salting room for salting bacon and making pickles; and another room for pastry, with a well-lit table for crimping pasties together or making ornate pie toppings. The existence of the pastry room reflected the aristocratic love of architectural raised pies and fancy tarts.

The least desirable room to work in would have been the scullery (hence the term scullery maid), which was the place for doing all the unpleasant pedestrian jobs: peeling vegetables, gutting fish, and washing dishes, no easy job when your only tools were boiling water, grimy rags, and soap. The scullery was dominated by a vast copper boiler for supplying wash water, and capacious stone sinks and plate-drying racks. It would have smelled of stale food and greasy suds. The floor needed to be angled so that the constant splashes of dirty water flowed off into a drain.

With the most unpleasant tasks confined to the scullery, the rich country house kitchen itself could be very pleasant. It was a room whose function was purely cooking, with none of the laundry work/dishwashing/food storage that we tend to cram into our kitchens. It tended to be a large stone-floored room—perhaps twenty by thirty feet—with expansive windows and whitewashed walls, dominated by a wooden kitchen table in the center on which were placed various work boards. Doors led off to the scullery and the larders. The kitchen would contain a dresser for utensils, and gleaming copper pots on shelves. There was plenty of room for the various cooks and cooks' maids to bustle about their business, cooking over the multiple heat sources, baking in the oven, making sauces over the stewing stove, steaming in a bain-marie, or roasting at the fire. When you visit a stately home and stand in such a kitchen, it is easy to feel a twinge of envy for all the spacious scrubbed wood, and to compare it with your own cramped galley kitchen at home and sigh. But sigh no more. These kitchens might be beautifully appointed, but they did not belong to the people doing the cooking: this was a place of work, not leisure. "Waste Not Want Not" was emblazoned on the walls of many of these kitchens, a reminder not to pilfer the food, because it was not yours. In the cities, Victorian kitchen drudges worked in more cramped conditions. The kitchen was often placed in a dank, beetle-infested basement, so that the unseemly business of cooking could be kept out of view, while the poor cooks sweltered unseen, hunching over their cast-iron kitcheners. Such Victorian kitchens had more in common with a professional restaurant kitchen than with our own home-cooking areas.

The great change of the twentieth century was the creation of new middle-class kitchens aimed at the people who were actually going to be doing the eating as well as the cooking. These new spaces were different from either the squalid one-room kitchen/living rooms of the preindustrial masses or the servant-run kitchens of the privileged. They were hygienic, floored with linoleum and powered by gas and electricity. The biggest change was that they were designed specifically around the needs of the people who used them. In 1893, Mrs. E. E. Kellogg (wife of the breakfast-cereal magnate) wrote that it was a "mistake" to think that just any room, "however small and unpleasantly situated is 'good enough' for a kitchen." Kellogg was part of a new "scientific" women's movement that sought to dignify the kitchen as a "household workshop."

The kitchen, thought Kellogg, was the key to the happiness of an entire family: it was the heart of a home. This idea is now so obvious to us, it is hard to register that it was not always so. Food has been a constant need in our lives, but the room from which it emanates has only existed in its current form since around the time of World War I. Humans have always cooked, but the concept of the "ideal kitchen" is a very modern invention.

The "kitchen of tomorrow" was a staple of twentieth-century life. There is a certain poignancy in looking back at photos of the futuristic kitchens of the past. You see people gazing in wonder at appliances that would now be considered rickety and archaic: tiny split-level electric ovens, miniscule fridges. Yesterday's future is to-morrow's junk. Or else you see that the vision of the future never took off. What seemed to be a new beginning was really a blind alley. One of the proudest exhibits at the British Ideal Home Exhibition in 1926 was a curious contraption consisting of a stovetop kettle with two saucepans joined on either side, so that three things could be cooked at once: a miracle energy-saving device that now looks like a joke. Gadgets that counted as futuristic on the eve of World War I included a thermos coffee pot that allowed coffee to be made hours in advance; potato ricers; revolving Lazy Susans (also known as Silent Waitresses because they saved women the trouble of waiting on their family at dinner); slaw cutters (mandolins for shredding raw cabbage); food choppers; cake mixers; ovens with glass doors to check the progress of food as it cooked; and above all, a heat source with modern fuel, whether kerosene, oil, or gas.

Yet for all these supposed labor savers, the main source of energy in most early twentieth-century kitchens continued to be that ex-pended by a single woman. Ideal kitchens were the product of a new servantless way of life for middle-class households. A succession of architects and home economics experts tried to devise a kitchen that would reduce the strain on women's bodies. In 1912, Christine Frederick, a writer for the *Ladies' Home Journal,* laid out plans through which the kitchen itself could become a labor- and time-saving device. Frederick became interested in the ideas of "scientific management" then in vogue in business. Efficiency engineers went into factories and advised on ways the workers could do the same work in less time. Why couldn't the same principles be applied in the kitchen? Frederick asked in her book, *The New Housekeeping.*

After a series of "home motion" studies made using real women of different heights, Frederick came up with an ideal kitchen design, arranged such that the worker using it performed the minimal number of steps, without ever stooping. Efficient cooking meant having the right tools grouped together before the task began, at the optimum level, and all utensils arranged with "proper regard to each other, and to other tasks." By arranging the kitchen in the most rational way possible, Frederick suggested, women could improve their efficiency by around 50 percent, freeing up time for other activities, whether reading, working, or "personal grooming." The right kitchen, Frederick argued, could buy women a little "individuality" and "higher life," though she does not suggest that men of the household might enjoy taking a turn at the stove; in 1912, that was a step too far.

Another rational kitchen of the early twentieth century was the Frankfurt Kitchen, created by the great Margarete Schütte-Lihotsky, the first female architecture student at the Vienna School of Arts and Crafts. Between 1926 and 1930, every apartment in the city of Frankfurt's public housing program was fitted with an identical kitchen built to Schütte-Lihotsky's specifications. Over a short space of time, more than 10,000 kitchens, all virtually indistinguishable, were built. These kitchens all had the same work surfaces and dish-drying racks, the same blue storage cabinets, the same waste disposal bin.

The Frankfurt Kitchen may have been small—though no smaller than many kitchens in modern-day New York City, where competitive complaining takes place over who has the tiniest galley kitchen—but it had a remarkable characteristic. It was based on how women actually moved around a kitchen, rather than on how a designer wanted them to behave.

In the 1920s kitchens of Britain and America, the all-purpose kitchen cabinet was promoted as a system for improving women's lives. These were forerunners of the built-in kitchen: systems of cupboards, shelves, and drawers, with swing-out work surfaces and containers for flour and sugar. Some even came equipped with built-in iceboxes. The biggest manufacturer was Hoosier, of Indiana. "Hoosiers," as they were called, took on the role of the kitchen dresser, the larder, and the kitchen table, all in one. "The Hoosier will help me to stay young," trumpeted an ad of 1919, showing a radiant newlywed bride.

The admen conjured up dream women to use these cabinets, but Hoosiers showed a lack of imagination about what real women needed in a kitchen. These cabinets were like toys for women to play in rather than serious work tools. By cramming everything into a single confined space, these isolated units made it much harder for anyone else in the family—children or husband—to help a woman with cooking or dish washing. They also prevented the cook from making full use of the room.

Compare this with the Frankfurt Kitchen, which came equipped with a swivel chair (height adjustable, a rare acknowledgment from an architect that humans come in different sizes) so that women could easily glide from the plain wooden work surface by the window to the cupboards and back again. Architects now speak of five basic arrangements for kitchens: L-shaped, U-shaped, island, one-wall, and galley. The toniest kitchens today are almost always laid out L-shaped or U-shaped, or around an island. But Schütte-Lihotsky showed what potential there is in the simple galley.

The kitchen's crowning glory was its storage system, resembling a filing cabinet in an office. There were fifteen aluminum drawers, arranged in three neat rows of five. Each one came inscribed with the name of a dry ingredient: flour, sugar, rice, dried peas, and so on. The drawers had sturdy handles, making them easy to lift out, one at a time, single-handed. The best part of the Frankfurt Kitchen was this: on the end of each drawer was a tapered scoop, so that whoever was cooking could lift out the drawer of rice, say, and pour the required amount—without spillage—straight onto the scales or into the pot. I have never seen such an ergonomically perfect solution to the storage of food as this one. It is beautiful, practical, time-saving, and systematic. It is all the more remarkable to find such a high quality of design in democratic kitchens made for working-class tenants.

Schütte-Lihotsky was a social revolutionary—the Nazis imprisoned her for four years for belonging to a Communist resistance group—and her kitchen had a feminist agenda. Schütte-Lihotsky was hopeful that the right kitchen design could help liberate women from their role as housewives, freeing up enough time so they could increasingly work outside the home. The Frankfurt tenants themselves, however, did not always feel liberated by their kitchens. Some disliked being forced to use electricity, complaining that an electric kitchen was expensive to run. Beyond that, they rebelled against the functional modernist aesthetic, and yearned for the clutter and mess of their old kitchens.

It took time for the brilliance of the Frankfurt Kitchen to be acknowledged. Schütte-Lihotsky's Communist beliefs meant that she did not get many commissions in her native Austria, even after the fall of Hitler. Finally, aged eighty-three, Schütte-Lihotsky was given the architecture prize of the city of Vienna. The Frankfurt Kitchen is now adored by architecture students, and it formed the centerpiece of an exhibition on kitchen technology at New York City's Museum of Modern Art in 2011. Walking around the exhibit, I saw New Yorkers, some of the most demanding consumers in the world, stop and stare admiringly at Schütte-Lihotsky's humble aluminum storage drawers. This was something that the postwar American kitchen, for all its plenty, did not have.

The Frankfurt Kitchen was a tiny galley, just a touch under 6.5 feet wide and almost 10 feet long. But the rational designers of the prewar years did not think that the ideal kitchen needed to be particularly spacious. Christine Frederick favored a room 10 by 12 feet, a little wider than the Frankfurt Kitchen but not much longer. Frederick knew that more space was a mixed blessing because it meant further for the person cooking to walk. The critical design factor was having tasks and equipment clustered together, encouraging a "chain of steps" around the room. Frederick identified six distinct stages in cooking a meal: preparing, cooking, serving, removing, washing, putting away. Each stage needed its own tools. At each stage, the tools should be at the right height and in the right position for the worker:

> Too often the utensils are all hung together, or jumbled in a drawer. Why reach across the stove for the potato masher when it belongs over the table? Why walk to the cabinet for the pancake turner when you need it for the stove?

Why indeed? Yet a hundred years later, it is striking how few of us manage to move around our kitchens with any real efficiency.

The problem is partly that Frederick's rational kitchen was not the only way to come up with an ideal kitchen. By the 1940s, this pragmatic approach to kitchen design had been supplanted by something much more elaborate: fancy cabinets, curvy ovens. Many ideal kitchens were—and are—less about introducing greater efficiency into the life you already live and more about pretending to lead another life altogether. We have chosen this room above all others as the one in which we will project a perfect vision of ourselves. For Frederick, the aim of the kitchen was "to see how few utensils, pots and pans are necessary." For most commercial kitchen designers, however, the aim has been to sell us as many lovely kitchen objects as possible; to replicate that sensation of mild, hyperventilating envy that we often feel on wandering around a kitchen showroom. How can life be complete without a built-in fuchsia pink bean-to-cup espresso machine?

From the 1940s onward, ideal kitchens were dangled above women's noses as a treat: a compensation for a life of drudgery or part of a sleight of hand that told them how lucky they were to be unpaid "homemakers." Christine Frederick's rational kitchen had been driven by efficiency: the fewest steps, the fewest utensils. The new ideal kitchens were far more opulent. These were dollhouses for grown women, packed with the maximum number of trinkets. The aim was not to save labor but to make the laborers forget they were working. As Betty Friedan wrote in *The Feminine Mystique*, the mid-century suburban kitchen started to take over the rest of the house. It was beautified with mosaics and vast purring fridges. Women were being encouraged—by advertisers, especially—to find emotional fulfillment in housework, to compensate for their lack of outside work. In 1930, 50 percent of American women were in paid work; by 1950 this had dropped to 34 percent (as against 60 percent in 2000).

The luxury of the midcentury kitchen was also a way of compensating for—or forgetting—the hardship of war. In 1944, the last year of war, the Libbey-Owens-Ford Glass Company created a "Kitchen of Tomorrow" that was seen by an estimated 1.6 million consumers across America.

Like most model kitchens, this prototype kitchen with glass Tufflex cupboards was intended to create jealousy—and therefore generate sales. *The Washington Post* noted that it offered a "bright" vision of a postwar future, which made it worth putting up with the present; the homemaker would "use what she has now cheerfully if she can aspire to such a kitchen after the war." America had it easy foodwise compared to any other country in the world during the war. But the perception within the country was still one of deprivation. To American women dealing with food rationing—notably of sugar and red meat—the sight of such a kitchen was a heady foretaste of the plenty to come.

Nearly seventy years later, this 1944 "Kitchen of Tomorrow" still looks remarkably high-tech in many ways; that is to say, it looks desirable. The floor paneling is dark and sleek, and there are cool glass backsplashes. The thing you notice most is that the designer— H. Albert Creston Doner—decided to do away with traditional pots and pans. In their place are a series of glass-topped vessels heated by electricity—a little like sous-vide baths—hidden under pedal-operated sliding panels. When not in use, the entire unit can be covered up to become a "study bench for the children or a bar for Dad." A spick-and-span 1940s model housewife sits at her pop-out sink conveniently located next to her pull-out vegetable drawer. She is peeling potatoes.

But here is where the high-tech vision collapses. The object with which this elegant woman peels her potatoes in this kitchen of tomorrow is just a plain old paring knife. This is not such a utopia after all. It may be a kitchen beyond pots and pans. But it lacks a decent peeler.

It's a small thing, but good vegetable peelers are a very recent development. They have been in our lives only since 1990. I count them among the most important technologies in the modern kitchen because these humble little tools have quietly made preparing a meal easier as well as subtly changing what and how we eat.

As I remember from growing up, peeling vegetables used to be one of the most annoying jobs in the kitchen. For centuries, the default method was to use a small paring knife with a tiny, pointy blade. In the right hands—those of a trained chef—the paring knife is an excellent tool, but it required immense concentration to remove every scrap of peel without also gouging your own thumb. If you weren't very adept with a paring knife, tough luck—there were no other options. The Sears Roebuck catalog for 1906—for many Americans the source from which all kitchen tools were bought—listed an apple corer and a wooden-handled paring knife, but no peeler.

By the mid-twentieth century, peelers were available, but they were bothersome to use in various ways. In Britain, the standard peeler was the Lancashire (named after one of the great potato-loving counties), with a handle tied around with string. The crude fixed blade was an extension of the handle. It was difficult to get any purchase on a potato or an apple without removing wasteful chunks of the flesh at the same time.

Far better were the swivel-action peelers of America and France, but these had drawbacks, too. The standard swivel peeler came with a handle of waffled chrome steel and a blade of carbon steel with a strip cut out of it. These sharp peelers were highly effective gadgets, because the blade contoured itself to the curve of the vegetables. But they hurt to use. As you put pressure on the vegetables, the sharp steel handle would correspondingly stab into the palm of your hand. Preparing mashed potatoes for a large family gathering could leave you with peeler-related blisters. Another option was the "Rex" style swivel peeler, whose curved metal handle was slightly more comfortable to hold, but it was—to my mind—even more difficult to use, because the shape of the handle forced you to pull clumsily away from yourself, instead of using the natural shucking motion of the standard swivel peeler.

In the late 1980s, Sam Farber's wife, Betsey, was finding it even harder than normal to peel vegetables, because of slight arthritis in her hand. A groundbreaking thought occurred to Farber, who had recently retired from the housewares business: why did peelers have to hurt? In designing a peeler that Betsey wouldn't find difficult to grip, Farber realized he could also make a tool that would be easy for everyone to hold. He approached the design firm Smart Design with his idea, and in 1990, after many prototypes were tried and rejected, the OXO vegetable peeler was unveiled at a gourmet show in San Francisco.

The OXO peeler represents a great piece of lateral thinking. You might assume that the way to build a better peeler would be to focus on the blade. Farber saw that the key for the cook using it was really the handle. The OXO's blade is super sharp, and angled like the old carbon-steel swivel peelers. What makes it different, however, is the soft, chunky, slightly ugly, black handle. It is made from Santoprene, a sturdy but softish compound of plastic and rubber, with little fins along the top, similar to those on bicycle handles, to absorb pressure. The shape is a fat oval, to prevent it from spinning in your hand. It feels comforting. It also really works, taking off just a paper-thin outer layer of peel. No matter how hard you press against a fruit or vegetable as you whisk off its skin, it doesn't hurt. With an OXO, you can even peel rock-hard butternut squashes, knobbly quinces, furry kiwifruits.

The OXO peeler was a game-changer. More than 10 million have been sold to date, and it threw open the entire market in peelers, causing numerous rivals to be invented: serrated fruit peelers, vegetable peelers with curved blades, Y-shaped and C-shaped and U-shaped peelers in any color you desire (and many you don't). Thirty years ago, if you went into a high-class kitchenware shop, you might see twenty different melon ballers—round cut, oval cut, double-ended, fluted, from large to small—but probably just two peelers, the Rex and the Lancashire. Peelers were regarded as hardware, a relic of the old scullery, when peeling vegetables was drudge work. Now the situation is reversed. Melon ballers have largely been banished from the kitchen, rejected as pretentious, whereas peelers are offered in every possible permutation. The owner of one UK kitchenware shop told me recently that he stocked sixty different peelers, allowing for all the color variations.

Peelers that work without discomfort are part of a new ergonomics in the kitchen. In the non-electrical utensils department, there are now ergonomic spatulas and ergonomic colanders, cushy soft-handled whisks and silicone basting brushes. Ergonomics is the study of designing equipment that fits the limitations and abilities of the human body. All kitchen tools should be ergonomic, because their role has—in theory—always been to help humans cook. But it is striking how often traditional designs have hindered our movement around a kitchen in small ways we do not even notice until a better way is shown. Until Microplane graters were launched in 1994—whose inspiration came when a Canadian housewife borrowed one of her husband's wood rasps to grate the zest for an orange cake—we accepted the fact that grating citrus was an inherently vexing task: it consisted of mangling a lemon on the smallest holes of a box grater, then desperately scraping off the meager strands of zest with a spoon. It turns out that all we needed was a better, sharper tool. Microplaned zest falls out with the easy grace of dandelion fluff.

Many of these ergonomic tools seem to bring us closer to the old preindustrial ways, when people tended to make their own tools. The household wooden spoon felt just right because it was whittled just for you. So much high-tech gadgetry is alienating because— however impressive it may be on its own terms—it seems to be doing battle with the human body. Ergonomic peelers and graters, on the other hand, are part of a new friendliness in kitchen objects: a willingness to address not just culinary problems, but problems in the way we experience cooking. Like the modernists, the designers of these new ergonomic tools approach the kitchen in a spirit of "why not?" The difference is that the aim of the question is not to reinvent cooking, but to make it easier.

For most cooks, ergonomics is a more useful way of thinking about the modern kitchen than whether something is high-tech or not. In the end, what you want from tools is not that they should be advanced but that they should work: to perform the job as helpfully as possible and fit with your particular kitchen and body, whether you are cooking for one, two, or many. In a single person's kitchen, this might mean one of those new boiling-water faucets (the Quooker) that enables you to make one portion of soothing pasta at the end of a long workday, very fast. For a big family, it might mean a steam oven, set on time delay to provide a tray of hot, nutritious food at a preordained time without any arguments over whose turn it is to make lunch. I recently visited a kitchen whose owners had tried to build everything on virtuous green principles, minimizing waste and carbon output. The work surfaces were all reclaimed; the German induction range was super-energy-efficient; the food to emerge from the eco pans was vegetarian. In contrast to the kitchens of the past, no one was exploited here; the cooking was shared fairly and evenly between the couple. The most inventive thing about the design of this room was also the simplest: they asked their carpenter to make their storage cupboard for cans and jars much shallower than normal, to stop them wasting food.

The most ergonomic tool for a given kitchen may or may not be the most newfangled. The kitchen island is a recent addition to our cooking lives, whose aim was to prevent the cook from facing the wall, but whose effect in many kitchens is to obstruct movement and imprison the person cooking behind the stove. To my mind, a kitchen table is a far more useful and sociable work surface. But you may disagree. Tools justify themselves—or not—through use. I know someone, a friend of my late grandmother, who has just given up on electric kettles—which most in Britain regard as an utterly indispensable technology—after one too many fuses blew. After decades of disappointing electric models, she had had enough and bought herself an old-fashioned stovetop whistling teakettle instead. It suits her better, she says. This is why, to answer Hervé This's question, we still cook with whisks, fire, and saucepans as they did in medieval times. We do so because most of the time, in most kitchens, whisks, fire, and saucepans still do the job pretty well. All we want is better whisks, better fire, and better saucepans.

Sometimes, you see mock-ups of historic kitchens. They might be part of an exhibition on ancient food or may be a bit of promotion for a kitchen appliance firm: a colorful look back at the history of our ovens! These mock-ups almost always make the same subtle mistake. It's not anachronism that's the problem: no Elizabethan televisions or 1920s computers. It's that these rooms are *too* authentic. Everything is made to fit the period in question; like a showroom kitchen, everything matches. A mock-up 1940s kitchen, for example, will include no item that wasn't made in the 1940s: there will be a 1940s toaster, 1940s pots and pans, a 1940s gas oven, a 1940s radio and 1940s chairs. Real kitchens aren't like that. In the kitchens we actually inhabit, old and new technologies overlap and coexist. A thirty-year-old housewife of 1940 would have had parents born in the nineteenth century; her grandparents would have been high Victorians, toasting bread by a grate with a fork; are we really to suppose that these earlier lives left no trace on her kitchen? No salamander? None of grandmother's cast-iron pans?

In the kitchen, old and new stand side by side as companions. In the grand kitchens of the past, when a new piece of equipment was adopted, it did not necessarily edge out the old. Successive tools were added on top, but the original ways of cooking could be glimpsed underneath, like a palimpsest.

Calke Abbey is an old Derbyshire house whose inhabitants, the Harpur family, hardly threw anything away. It now belongs to the National Trust and remains in a state of considerable decrepitude. The large old kitchen is really a series of kitchens, one on top of the other, each representing a slice of time. This stone-flagged room was first fitted as a kitchen in 1794 (before that, it may have been a chapel). The kitchen clock was bought in Derby that year. Also original to 1794 is a vast old roasting hearth, with a clockwork spit-jack on top. In front of this fire, beef would once have turned on its spits. But sometime in the 1840s, roasting must have been abandoned, as a closed-off cast-iron oven was shoved into this hearth. Later, this oven, too, must have failed to meet the household's needs, for in 1889, a second hearth was added with an additional cast-iron stove. Meanwhile, along another wall there is an eighteenth-century-style stewing stove set in brickwork, used for stewing and sauces. Finally, in the 1920s the inhabitants installed a modern Beeston boiler for hot water alongside the old ranges. At no stage did anyone think to remove any of the previous cooking tools. In 1928, with the number of servants in the house suddenly reduced, the room was abruptly abandoned; a new, more functional kitchen was set up elsewhere in the house. The old kitchen remains now as it was in 1928. A dresser still stands, filled with rusting pots and pans. The spit-jack and the kitchen clock still hang on the walls, just where they were first placed.

Of course, most households are more ruthless about discarding things when they fall out of use. But kitchens remain extremely good at accommodating both old and new under a single roof. There is something sad as well as wasteful about the current impulse to start a kitchen from scratch: to rip out every trace of the cooks who came before you. It feels forgetful. Kitchens in general have never been so highly designed; so well equipped; so stylish; or so soulless. In the 1910s, the ideal was the "rational" kitchen; later, in the 1940s and 1950s, it was the "beautiful" kitchen. Now, it is the "perfect" kitchen. Everything must match and fit, from ivory ceiling to limestone floor. Every element must be "contemporary." Anything shabby or out of place is discarded (unless you've gone for "shabby chic" as your vibe).

It's an illusion, of course. In the most highly designed modern kitchen, we are still drawing on the tools and techniques of the past. As you grasp your shiny tongs to whip up a modern dish of wok-fired squid and greens or linguini with butternut squash and red chili, you are still doing an old, old thing: using the transformative power of fire to make something taste better. Our kitchens are filled with ghosts. You may not see them, but you could not cook as you do without their ingenuity: the potters who first enabled us to boil and stew; the knife forgers; the resourceful engineers who designed the first refrigerators; the pioneers of gas and electric ovens; the scale makers; the inventors of eggbeaters and peelers.

The food we cook is not only an assemblage of ingredients. It is the product of technologies, past and present. One sunny day, I decide to make a quick omelette for my lunch, a puffy golden oval in the French rolled tradition. On paper, it consists of nothing but eggs (free range); sweet, cold butter; and sea salt, but the true components are many more. There is the fridge from which I fetch the butter and the old battered aluminium frying pan in which I cook it, whose surface is seasoned from ten years of use. There is the balloon whisk that beats the eggs, though a fork would do just as well. The countless cookery writers whose words warned me not to *over*beat. The gas burner that enables me to get the pan hot enough but not so hot the eggs burn or get rubbery. The spatula that rolls the golden-brown omelette onto the plate. Thanks to all these technologies, the omelette has on this occasion, for this particular solitary lunch, worked. I am pleased. The entire mood of an afternoon can be spoiled or improved by lunch.

There is still one more component to this meal, however: the impulse to make it in the first place. Kitchens only come alive when you cook in them. What really drives technology is the desire to use it. This omelette lunch would never have been made without my mother, who first taught me that the kitchen was a place where good things happen.

~ *Coffee* ~

COFFEE TECHNOLOGY HAS BECOME PERPLEXING.
The inventiveness lavished on this substance reflects its status as the world's culinary drug of
choice. To brew coffee is to do nothing more than
mix grounds with hot water and strain out the
dregs. But methods for doing this have varied
wildly, from the Turkish ibriks used to make rich
dark coffee since the sixteenth century to the my-
pressi TWIST launched in 2008, a handheld
espresso machine powered with gas canisters like a
cream whipper.

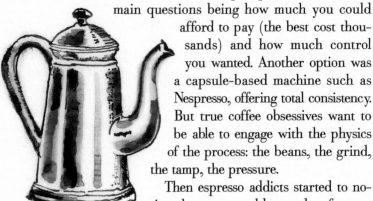

Only a couple of years ago, the last word in
coffee makers was the huge espresso machine, the
main questions being how much you could
afford to pay (the best cost thousands) and how much control
you wanted. Another option was
a capsule-based machine such as
Nespresso, offering total consistency.
But true coffee obsessives want to
be able to engage with the physics
of the process: the beans, the grind,
the tamp, the pressure.

Then espresso addicts started to notice that you could spend a fortune
and do everything right and still end up with
mediocre coffee—there were just too many variables. The new wave of coffee technology has
moved beyond espresso machines—indeed, largely
beyond electricity. There's the AeroPress, a clever
plastic tool that uses air pressure to force coffee
down a tube into a mug. All you need is a kettle

and strong arms. Still trendier is the Japanese siphon. It looks like something from a chemistry class: two interconnected glass bulbs with a small burner underneath. But people of a certain age point out that these siphons are not so different from the Cona coffee maker of the 1960s.

The real action in coffee now is low-tech. We've spent so long thinking about ways to make better coffee, we've come full circle. The most avant-garde coffee experts in the world—in London, Melbourne, and Auckland—now favor French press and filter over pricey espresso machines. It's only a matter of time before someone announces the next big thing: the pitcher and spoon.

ACKNOWLEDGMENTS

The epigraph to Chapter 7, "This Is Just to Say," by William Carlos Williams, is taken from *The Collected Poems: Volume I, 1909–1939,* copyright © 1938 by New Directions Publishing Corp. Reprinted by permission of New Directions Publishing Corp.

The person I really want to thank is the great Pat Kavanagh, who died in 2008. I will always be grateful that she was my agent. It was Pat who led me to Helen Conford at Penguin, whose idea this book was, and who has been the most conscientious and insightful editor I could wish for; Helen disproves the adage that no one really edits books any more. Also at Penguin, among others, I'd like to thank Patrick Loughran, Penelope Vogler, Lisa Simmonds, and Jane Robertson.

Pat also brought me to my two superb agents, Sarah Ballard at United Agents in London and Zoe Pagnamenta at the Zoe Pagnamenta Literary Agency in New York; I am so thankful to them both; also to Lara Hughes-Young at United Agents.

I owe huge thanks, too, to Lara Heimert at Basic Books, for her patience, encouragement, and intelligent editorial judgment. Also at Basic, thanks to Katy O'Donnell, Michele Jacob, Caitlin Graf, Michelle Welsh-Horst, Cisca Schreefel, and Michele Wynn, to whom I am particularly grateful for her careful copyediting.

Annabel Lee produced wonderful illustrations at very short notice; I wish that my own kitchen implements looked half as good. Carolin Young kindly read the book with a food historian's eye, but needless to say, any mistakes that remain are my own. At an early stage of writing, I took part in an edition of the BBC Radio 4 Food Programme on gadgets, which was a great help in refining some of my ideas; thanks so much to Sheila Dillon and Dilly Barlow. I am also grateful to the editor of my food column in *Stella* magazine, Elfreda Pownall. All my love and thanks to my family—David, Tom, Tasha, and Leo—for putting up with the curious new gadgets coming into the house and the dull visits to stately home kitchens; and special thanks to Tom for the title ideas (even if we didn't use yours in the end).

Much of the research was done at the Cambridge University Library; and at the Australian National University in Canberra (thanks to Bob Goodin). Finally, for help, advice, or assistance of various kinds I would like to thank, among others: Amy Bryant, Catherine Blyth, David Burnett, Sally Butcher, John Cadieux, Melissa Calaresu, Tracy Calow, Ivan Day, Katie Drummond, Katherine Duncan-Jones, Gonzalo Gil, Sophie Hannah, Claire Hughes, Tristram Hunt, Tom Jaine, Beeban Kidron, Miranda Landgraf, Frederika Latif at John Lewis, Reg Lee, Anne Malcolm, Esther McNeill, Anthea Morrison, Anna Murphy, John Osepchuk, Kate Peters, Ben Phillips at Steamer Trading, Sarah Ray, Miri Rubin, Cathy Runciman, Lisa Runciman, Ruth Runciman, Garry Runciman, Helen Saberi, Abby Scott, Benah Shah at OXO, Gareth Stedman Jones, Alex Tennant at Aerobie, Robert and Isabelle Tombs, Mark Turner, Robin Weir, Andrew Wilson, and Emily Wilson.

NOTES

GENERAL NOTES ON SOURCES

In a book covering so much ground, I inevitably owe a great deal to many second-ary sources, from journal articles to chapters and books, in addition to the primary sources I consulted, from historic cookbooks to works on technology, to contem-porary newspapers and other periodicals, to catalogs of kitchenware such as Sears, Roebuck for the United States and Jacquotot for France; and to the kitchens I vis-ited. The bibliography gives a fuller list of sources consulted, but the notes here will single out those that were especially helpful.

When I was first starting to think about this subject, a friend gave me Molly Harrison's *The Kitchen in History* (Osprey 1972) and it remained a useful refer-ence point throughout. I am also indebted to *Irons in the Fire: A History of Cooking Equipment* (Crowood Press 1984) by Rachel Feild, who approaches the subject of kitchen tools from an antiquarian perspective.

Anyone who is remotely interested in the history of food should read Reay Tannahill's wonderful essay *Food in History* (updated edition, 2002). On cooks in history, *A History of Cooks and Cooking* (Prospect Books 2001) by Michael Symons is both provocative and packed with information. Another panoramic overview is *Food: A History* by Felipe Fernández-Armesto (Macmillan 2001).

I am grateful to the Oxford Symposium on Food and Cookery, an annual gath-ering cofounded by Alan Davidson and Theodore Zeldin, which remains one of the best occasions for the study and appreciation of food in history. There are countless fascinating nuggets in the Symposium Proceedings, published each year by Prospect Books, which also publishes *Petits Propos Culinaires*, an invaluable journal for food historians (and despite the title, not in French). Another great periodical on food history is *Gastronomica,* edited by Darra Goldstein. I also owe a debt to Ivan Day and Peter Brears, two remarkable food historians whose work, often through the Leeds History of Food Symposium, has been unusual for its emphasis on the techniques and equipment of historic cookery.

Among the general books I have found most useful on cooking technology, seen in context as one aspect of domestic life in Britain, I strongly recommend Caroline Davidson's superb *A Woman's Work Is Never Done: A History of Housework in the British Isles, 1650–1950* (Chatto & Windus 1982) and Christina Hardyment's *From Mangle to Microwave: The Mechanisation of Household Work* (Polity 1990); the latter covers the modern period up to 1990. For the American side of the same story, told from a feminist point of view, *More Work for Mother* (Basic Books 1983) by Ruth Schwartz Cowan is thought-provoking. All three of these are brilliant works of social history, as much as histories of gadgetry.

There are countless fine guides to modern kitchen tools. The one I returned to most was James Beard's encyclopedic work *The Cooks' Catalogue* (New York 1975): not for nothing is he still remembered as one of the great American food writers. His combination of knowledge and passion makes him always worth reading. Also useful is Burt Wolf's updated version of the same work: *The New Cooks' Catalogue* (Alfred Knopf 2000), a fine guide to everything from pastry knives to food processors. For something more up-to-date, I like *Alton Brown's Gear for Your Kitchen* (Stewart Tabori Chang 2008); or for the futuristic kitchen, Jeff Potter's exciting *Cooking for Geeks: Real Science, Great Hacks, and Good Food* (O'Reilly Media 2010), which will tell you everything from how to improvise your own sous-vide machine to a method for cooking salmon in the dishwasher.

Introduction

On the dying tradition of wooden eating spoons, see Rogers (1997).

For examples of traditional histories of technology that pay little or no attention to food, see Larson (1961), which covers neither food nor cooking; Derry and Williams (1960), which covers the plough and the threshing machine but not kitchen tools; Forbes (1950), which includes canning but not domestic food technology.

Linda C. Brewster's debittering patents are among the many inventions by women listed in Stanley (1993).

On the link between pottery and the survival of the toothless, see Brace (2000) and Brace et al. (1987).

The idea of the hidden intelligence of tools is explored in the brilliant Weber (1992).

On frigophobia at Les Halles, see Claflin (2008).

The 2011 survey of British cooking habits was commissioned by 5by25, a campaign aimed at getting people to learn five dishes by the age of twenty-five.

The Japanese research on feeding rats differently textured pellets is written up in Oka et al. (2003).

The revolution of brick chimney cooking is explored in Feild (1984), which also mentions the irony of cans being invented so long before can-openers.

Kranzberg's laws of technology are laid out in Kranzberg (1986).

Cowan's arguments about communal cooking are in Cowan (1983).

CHAPTER ONE: POTS AND PANS

By far the most useful source for writing this chapter was Jaine (1989), the Proceedings of the 1988 Oxford Symposium on the Cooking Pot. This volume includes Lemme (1989) on the ideal pot, Gordon and Jacobs-McCusker on one-pot cookery, and Coe on the Maya cooking pot, among many other excellent essays. The anthropological and archaeological literature on early pottery is immense. On the origins of pottery, see, for example, the Hoopes and Barnett volume (1995), Arnold (1985), Childe (1936), Pierce (2005). On Jomon pottery, see Aikens (1995). On turtle cookery, see Man (1932) and Bates (1873). On pit ovens, see Wandsnider (1997), Doerper and Collins (1989), Thoms (2009), and many more. On Greek pottery, see Vitelli (1989) and (1999), Sparkes (1962), Soyer (1853). On cauldrons, see Feild (1984), Wheaton (1983), Brannon (1984). On the *batterie de cuisine,* see Brears and Sambrook (1996).

The nineteenth-century authors I consulted on the question of boiling included Beeton (2000), Blot (1868), Buchanan (1815), Kitchiner (1829).

For a great account of the shortcomings of cooking with nonstick, and much else besides, see Harris (1980).

On the relationship between cooking pots and burners, see Myhrvold (2011).

CHAPTER TWO: KNIFE

On Stone Age technologies of cutting, see Toth and Schick (2009), Davidson and McGrew (2005), and Wrangham (1999).

On the history of the tou and its contribution to Chinese cuisine, see Chang (1977), particularly the essays by Anderson and Anderson and by Chang himself; also Symons (2001); and Book 10 of the Analects of Confucius.

For a practical guide to Chinese methods of cutting and the tou (as well as sumptuous recipes for what to do with your tou-chopped ingredients), see Dunlop (2001); also Dunlop (2004), an article on the tou. Dunlop is the Elizabeth David of Chinese food, one of the great food writers today.

On European carving, see Furnivall (1868), Worde (2003), Brears (1999) and Brears (2008); also Visser (1991). On the Sheffield cutlery industry, see Lloyd (1913). On European knives as part of European civilization, see Marquardt (1997) and all the essays in Coffin et al. (2006).

The complete court proceedings of the case of Joseph Baretti and his prosecution for murder by fruit knife can be read at www.oldbaileyonline.org/; I recommend it; the trial makes scintillating reading.

Charles Loring Brace is a prolific scholar; among the many papers in which he has set out his thesis about the overbite and other aspects of human teeth are Brace (1977), Brace (1984), Brace (1986), Brace et al. (1987), Brace (2000). Ferrie (1997) is an interview with Brace in which he discusses how his interest in teeth developed. Also on Brace and his career, see Falk and Seguchi (2006) by two of his former pupils.

For practical appreciations of the joy of knives, which to buy and how to use them, see Jay (2008), Hertzmann (2007), McEvedy (2011). My own favorite knife is carbon steel with a rosewood handle and comes from Wildfire Cutlery in Oregon; I am grateful to McEvedy for the recommendation.

CHAPTER THREE: FIRE

For a sense of Ivan Day and his work, see www.historiccookery.com; also Day (2000), Day (2009), and Day (2011). Where Day is quoted here, it is mostly based on conversations with the author.

On cooking as the moment that made us human, see Wrangham (2009) and Wrangham, Jones et al. (1999). On the technology of open hearth cookery, Day (2009), Brears (2009). Eveleigh (1991) is one of the best sources on the English tradition of roasting before an open fire; see also Eveleigh (1986) on roasting utensils. Rogers (2003) is good on the English affection for roast beef. On the risks of fire in premodern times, see Feild (1984) and Hanawalt (1986). For Rumford's theory of open hearths versus closed stoves, see Rumford (1968), vol. 3. On early bread ovens as studied by archaeologists, see Waines (1987) and Samuel (1999). For an extraordinary report on the impact of smokeless stoves in the developing world, see Bilger (2009).

On gas ovens, see Young (1897), Webster (1844), Sugg (1890), Ohren (1871), Hardyment (1990), and Davidson (1982). On microwave technology, see Osepchuk (1984) and Osepchuk (2010) and Buffler (1993). On the culinary potential of microwave ovens, see Kafka (1987); also Myhrvold (2011), which includes a series of don't-try-this-at-home experiments for microwaves.

Fernández-Armesto's attack on microwaves appears in Fernández-Armesto (2001).

CHAPTER FOUR: MEASURE

This chapter was initially inspired by Sokolov (1989), a typically brilliant and provocative essay on America's system of cup measuring. For general histories of measurement, not confined to the kitchen, see Sydenham (1979), Tavernor (2007),

Whitelaw (1997). On Fannie Farmer, see the excellent Shapiro (1986), Smith (2007), the entry on Farmer by Harvey Levenstein in American National Biography Online, and Farmer herself: Farmer (1896) and Farmer (1904). Some other cookbook authors I consulted on measurement included but were not limited to: Blot (1868), Ellet (1857), Lincoln (1884), Martino (2005), May (2000), Parloa (1882), Rorer (1902). For Molokhovets, see Toomre (1992). Elizabeth David's acerbic comment on measuring comes in David (1970). For modernist measurements, see Blumenthal (2009) and Myhrvold (2011). Judy Rodgers's wise words on measuring appear on pp. 40 and 41 of Rodgers (2002).

Chapter Five: Grind

The medieval pancake recipe is in Power (1992).

On early grinding tools, see Wright (1994), MacGregor (2010), Ebeling and Rowan (2004). On the mortar and pestle, see Rios (1989).

On the Elizabethan love of beaten egg whites, see Spurling (1986) and Wilson (1973).

On labor-saving devices, and the lack of them before modern times, see Yarwood (1981). On the profusion of labor-saving devices as a mixed blessing, see Hardyment (1990). On the American eggbeater bubble, see Harland (1873) and Thornton (1994).

On food processors, see Collins (1989), De Groot (1977), David (1998), Barry (1983), Claiborne (1976), Claiborne (1981), Bittman (2010). On making kibbé, see Helou (2008).

Chapter Six: Eat

On spoons, see especially Emery (1976); also Snodin (1974), Homer (1975), Hughes and Hughes (1952). On cutlery in general, see Visser (1991) and Coffin et al. (2006), especially the essays by Young (2006) and Goldstein (2006). The etiquette books I consulted included Troubridge (1926), Ordway (1918), Green (1922), Anonymous (1836), Post (1960). On forks, see Young (2006), Coryate (1978), Serventi and Saban (2002), chap. 2, and Elias (1994). On chopsticks, see Barthes (1982), Chang (1977), Coe (2009), Hosking (1996), Ishige (2001). Bill Clinton's speech about sporks is available at http://www.c-spanvideo.org/program/63940–1, accessed July 2011. Also on sporks, see Koerner (2006), Lawrence (2010), and www.spork.org.

Chapter Seven: Ice

Much has been written about the Kitchen Debate. See, for example, Reid (2002), Reid (2005), Reid (2009), Oldenziel and Zachmann (2009), Larner (1986).

For a contemporary account, see Salisbury (1959). On the history of ice, see David (1994a), Beckmann (1817), Masters (1844). The greatest book on America and refrigeration is the panoramic and phenomenally scholarly Anderson (1953). On Mrs. Marshall and ice cream, see Weir, Brears, Deith, and Barham (1998); Weir and Weir (2010); Day (2011); as well as Marshall (1857), Marshall (1894), and Marshall (1896). On canned and frozen food, see Shephard (2000) and Friedberg (2009). On Einstein's fridge, see Segre (2002). On gas fridges, see Cowan in MacKenzie and Wajcman (1985). On various aspects of twentieth-century refrigerators, see Hardyment (1990), Nickles (2002), Plante (1995), Isenstadt (1998). On the frozen-food business, see Haddock (1954), Hamilton (1955); I also consulted various newspaper archives on this subject and the history of ice.

Chapter Eight: Kitchen

Hervé This's views on kitchen technology are expressed in This (2005) and This (2009).

On space food, see Levi in Friedland (2009). On the brief history and practice of sous-vide, see Hesser (2005), Keller (2008), Myhrvold (2011), also Renton (2010). On Kurti, see Kurti and Kurti (1988) and Sanders (2000). For general histories of the kitchen as a space, see Harrison (1972), Yarwood (1981). The radio program featuring Myhrvold and Waters was the excellent Freakonomics podcast, "Waiter, There's a Physicist in My Soup!" Part 1 first broadcast January 26, 2011. For Frederick, see Frederick (1916). On the Frankfurt Kitchen, see Kinchin (2011), Hessler (2009), Oldenziel and Zachmann (2009).

BIBLIOGRAPHY

Abend, Lisa (2011). *The Sorcerer's Apprentices: A Season in the Kitchen at Ferran Adrià's Elbulli*. New York, Free Press.

Aikens, Melvin (1995). "First in the World: The Jomon Pottery of Early Japan," in *The Emergence of Pottery: Technology and Innovation in Ancient Societies*, edited by William K. Barnett and John W. Hoopes, pp. 11–21. Washington, DC, Smithsonian Institution Press.

Akioka, Yoshio (1979). *Japanese Spoons and Ladles*. Tokyo, New York, Kodansha International.

Anderson, Oscar Edward (1953). *Refrigeration in America: A History of a New Technology and Its Impact*. Princeton, Princeton University Press.

Anonymous (1836). *The Laws of Etiquette by A Gentleman*. Philadelphia, Carey, Lea and Blanchard.

Appert, Nicolas (1812). *The Art of Preserving All Kinds of Animal and Vegetable Substances for Several Years*. London, Black, Parry and Kingsbury,

Arnold, Dean E. (1985). *Ceramic Theory and Cultural Process*. Cambridge, Cambridge University Press.

Artus, Thomas (1996). *L'Isle des Hermaphrodites*, edited by Claude-Gilbert Dubois. Geneva, Droz.

Artusi, Pellegrino (2004). *Science in the Kitchen and the Art of Eating Well*, foreword by Michele Scicolone, translated by Murtha Baca and Stephen Sartarelli. Toronto, University of Toronto Press.

Bailey, Flora L. (1940). "Navaho Foods and Cooking Methods." *American Anthropologist*, vol. 42, pp. 270–290.

Bang, Rameshwar L., Ebhrahim, Mohammed K., and Sharma, Prem N. (1997). "Scalds Among Children in Kuwait." *European Journal of Epidemiology*, vol. 13, pp. 33–39.

Barham, Peter (2001). *The Science of Cooking*. Berlin, London, Springer.

Barley, Nigel (1994). *Smashing Pots: Feats of Clay from Africa*. London, British Museum Press.

Barry, Michael (1983). *Food Processor Cookery*. Isleworth, ICTC Ltd.

Barthes, Roland (1982). *Empire of Signs*, translated by Richard Howard. London, Cape.

Bates, Henry Walter (1873). *The Naturalist on the River Amazon*, 3rd ed. London, John Murray.

Beard, James, ed. (1975). *The Cooks' Catalogue*. New York, Harper & Row.

Beard, Mary (2008). *Pompeii: The Life of a Roman Town*. London, Profile.

Beckmann, John (1817). *A History of Inventions and Discoveries*, 3rd ed. in 4 vols. London, Longman, Hurst, Rees, Orme and Brown.

Beeton, Isabella (2000). *The Book of Household Management, A Facsimile of the 1861 Edition*. London, Cassell.

Beier, Georgina (1980). "Yoruba Pottery." *African Arts,* vol. 13, pp. 48–52.

Beveridge, Peter (1869). "Aboriginal Ovens." *Journal of the Anthropological Society of London*, vol. 7, pp. clxxxvi–clxxxix.

Bilger, Burkhard (2009). "Hearth Surgery: The Quest for a Stove That Can Save the World." *New Yorker*, December 21.

Birmingham, Judy (1975). "Traditional Potters of the Kathmandu Valley: An Ethnoarchaeological Study." *Man*, New Series, vol. 10, no. 3, pp. 370–386.

Bittman, Mark (2010). "The Food Processor: A Virtuoso One-Man Band." *New York Times*, September 14.

Blot, Pierre (1868). *Handbook of Practical Cookery for Ladies and Professional Cooks Containing the Whole Science and Art of Preparing Human Food*. New York, D. Appleton.

Blumenthal, Heston (2009). *The Fat Duck Cookbook*. London, Bloomsbury.

Boardman, Brenda, Lane, Kevin, et al. (1997). *Decade: Transforming the UK Cold Market*. Energy and Environment Programme, University of Oxford.

Bon, Ottaviano (1653). *A Description of the Grand Seignour's Seraglio, or Turkish Emperor's Court*. London, Jo. Marton and Jo. Ridley.

Booker, Susan M. (2000). "Innovative Technologies: Chinese Fridges Keep Food and the Planet Cool." *Environmental Health Perspectives*, vol. 108, no. 4, p. A164.

Bottero, Jean (2004). *The Oldest Cuisine in the World: Cooking in Mesopotamia*. Chicago, University of Chicago Press.

Brace, C. Loring (1977). "Occlusion to the Anthropological Eye," in *The Biology of Occlusal Development,* edited by James McNamara, pp. 179–209. Ann Arbor, Center for Human Growth and Development.

—— (1986). "Egg on the Face, *f* in the Mouth and the Overbite." *American Anthropologist*, New Series, vol. 88, no. 3, pp. 695–697.

—— (2000). "What Big Teeth You Had, Grandma!" in C. Loring Brace, *Evolution in an Anthropological View*, pp. 165–199. Walnut Creek, CA, AltaMira.

——, with Rosenberg, Karen R., and Hunt, Kevin D. (1987). "Gradual Change in Human Tooth Size in the Late Pleistocene and Post-Pleistocene." *Evolution*, vol. 41, no. 4, pp. 705–720.

——, with Shao Xiang-Qing and Zhang Z. B. (1984). "Prehistoric and Modern Tooth Size in China," in *The Origins of Modern Humans: A World Survey of the Fossil Evidence*, edited by F. H. Smith and F. Spencer. New York, A. R. Liss.

Brannon, N. F. (1984). "An Examination of a Bronze Cauldron from Raffrey Bog, County Down." *Journal of Irish Archaeology*, vol. 2, pp. 51–57.

Brears, Peter (1999). *All the King's Cooks*. London, Souvenir Press.

—— (2008). *Cooking and Dining in Medieval England*. Totnes, Prospect Books.

—— (2009) "The Roast Beef of Windsor Castle," in *Over a Red-Hot Stove: Essays in Early Cooking Technology*, edited by Ivan Day. Totnes, Prospect Books.

——, and Sambrook, Pamela, eds. (1996). *Country House Kitchen, 1650–1900: Skills and Equipment for Food Provisioning*. Stroud, Alan Sutton (for the National Trust).

Brown, Alton (2008). *Alton Brown's Gear for Your Kitchen*. New York, Stewart Tabori Chang.

Buchanan, Robertson (1815). *A Treatise on the Economy of Fuel*. Glasgow, Brash & Reid.

Buffler, Charles R. (1993). *Microwave Cooking and Processing: Engineering Fundamentals for the Food Scientist*. New York, Van Nostrand Reinhold.

Bull, J. P., Jackson, D. M., and Walton, Cynthia (1964). "Causes and Prevention of Domestic Burning Accidents." *British Medical Journal*, vol. 2, no. 5422, pp. 1421–1427.

Burnett, John (1979). *Plenty and Want: A Social History of Diet in England from 1815 to the Present Day*. London, Scolar Press.

—— (2004). *England Eats Out: A Social History of Eating Out in England from 1830 to the Present*. London, Pearson Longman.

Bury, Charlotte Campbell (1844). *The Lady's Own Cookery Book*, 3rd ed. London, Henry Colburn.

Chang, K. C., ed. (1977). *Food in Chinese Culture: Anthropological and Historical Perspectives*. New Haven, Yale University Press.

Child, Julia (2009). *Mastering the Art of French Cooking*. London, Penguin.

Childe, V. Gordon (1936). *Man Makes Himself*. London, Watts.

Claflin, Kyri Watson (2008). "Les Halles and the Moral Market: Frigophobia Strikes in the Belly of Paris," in *Food and Morality: Proceedings of the Oxford Symposium on Food and Cookery 2007*, edited by Susan R. Friedland. Totnes, Prospect Books.

Claiborne, Craig (1976). "She Demonstrates Hour to Cook Best with New Cuisinart." *New York Times*, January 7.

———— (1981). "Mastering the Mini Dumpling." *New York Times*, June 21.

Clarke, Samuel (1670). *A True and Faithful Account of the Four Chiefest Plantations of the English in America to Wit, of Virginia, New-England, Bermudas, Barbados.* London, Robert Cavel.

Codrington, F. I. (1929). *Chopsticks.* London, Society for Promoting Christian Knowledge.

Coe, Andrew (2009). *Chop Suey: A Cultural History of Chinese Food in the United States.* Oxford, Oxford University Press.

Coe, Sophie D. (1989). "The Maya Chocolate Pot and Its Descendants," in *Oxford Symposium on Food and Cookery 1988, The Cooking Pot: Proceedings*, pp. 15–22. Totnes, Prospect Books.

Coffin, Sarah D., Lupton, Ellen, Goldstein, Darra, and Bloemink, Barbara, eds. (2006). *Feeding Desire: Design and the Tools of the Table.* New York, Assouline, in collaboration with Smithsonian Cooper-Hewitt.

Coles, Richard, McDowell, Derek, and Kirwan, Mark J., eds. (2003). *Food Packaging Technology.* Oxford, Blackwell.

Collins, Shirley (1989). "Getting a Handle on Pots and Pans," in *Oxford Symposium on Food and Cookery 1988, The Cooking Pot: Proceedings*, pp. 22–28. Totnes, Prospect Books.

Cooper, Joseph (1654). *The Art of Cookery Refined and Augmented.* London, R. Lowndes.

Coryate, Thomas (1978 [first published 1611]). *Coryats Crudities Hastily Gobled Up in Five Months Travells in France, Savoy, Italy.* London, William Stansby.

Cowan, Ruth Schwartz (1983). *More Work for Mother: The Ironies of Household Technology from the Open Hearth to the Microwave.* New York, Basic Books.

Cowen, Ruth (2006). *Relish: The Extraordinary Life of Alexis Soyer.* London, Weidenfeld & Nicholson.

Dalby, Andrew, and Grainger, Sally (1996). *The Classical Cookbook.* London, British Museum Press.

Darby, William, Ghalioungui, Paul, and Grivetti, Louis (1977). *Food: The Gift of Osiris.* London, Academic Press.

David, Elizabeth (1970). *Spices, Salt and Aromatics in the English Kitchen.* Harmondsworth, Penguin.

———— (1994a). *Harvest of the Cold Months: The Social History of Ice and Ices.* London, Michael Joseph.

———— (1994b [first published London 1977]). *English Bread and Yeast Cookery,* new American ed. Newton, MA, Biscuit Books Inc.

———— (1998 [first published 1960]). *French Provincial Cooking.* London, Penguin.

Davidson, Caroline (1982). *A Woman's Work Is Never Done: A History of Housework in the British Isles, 1650–1950.* London, Chatto & Windus.

Davidson, I., and McGrew, W. C. (2005). "Stone Tools and the Uniqueness of Human Culture." *Journal of Royal Anthropological Institute*, vol. 11, no. 4, December, pp. 793–817.

Day, Ivan (2011). *Ice Cream: A History*. Oxford, Shire Publications.

———, ed. (2000). *Eat, Drink, and Be Merry: The British at Table, 1600–2000*. London, Philip Wilson Publishers.

———, ed. (2009). *Over a Red-Hot Stove: Essays in Early Cooking Technology*. Totnes, Prospect Books.

De Groot, Roy Andries (1977). *Cooking with the Cuisinart Food Processor*. New York, McGraw-Hill.

De Haan, David (1977). *Antique Household Gadgets and Appliances, c.1860 to 1930*. Poole, Blandford Press.

Deighton, Len (1979). *Basic French Cooking* (revised and enlarged from *Où est le garlic?*). London, Jonathan Cape.

Dench, Emma (2010). "When Rome Conquered Italy." *London Review of Books*, February 25.

Derry, T. K., and Williams, Trevor I. (1960). *A Short History of Technology from the Earliest Times to A.D. 1900*. Oxford, Clarendon Press.

Doerper, John, and Collins, Alf (1989). "Pacific Northwest Indian Cooking Vessels," in *Oxford Symposium on Food and Cookery 1988, The Cooking Pot: Proceedings*, pp. 28–44. Totnes, Prospect Books.

Dubois, Urbain (1870). *Artistic Cookery: A Practical System Suited for the Use of Nobility and Gentry and for Public Entertainments*. London.

Dugdale, William (1666). *Origines Juridiciales, or Historical Memorials of the English Laws*. London, Warren.

Dunlop, Fuchsia (2001). *Sichuan Cookery*. London, Penguin Books.

——— (2004). "Cutting It Is More Than Cutting Edge." *Financial Times*, August 7.

Eaton, Mary (1823). *The Cook and Housekeeper's Complete and Universal Dictionary*. Bungay, J. and R. Childs.

Ebeling, Jennie (2002). "Why Are Ground Stone Tools Found in Middle and Late Bronze Age Burials?" *Near Eastern Archaeology*, vol. 65, no. 2, pp. 149–151.

Ebeling, Jennie R., and Rowan, Yorke M. (2004). "The Archaeology of the Daily Grind: Ground Stone Tools and Food Production in the Southern Levant." *Near Eastern Archaeology*, vol. 67, no. 2, pp. 108–117.

Edgerton, David (2008). *The Shock of the Old: Technology and Global History Since 1900*. London, Profile.

Elias, Norbert (1994 [first published 1939]). *The Civilising Process*, translated by Edmund Jephcott. Oxford, Blackwell.

Ellet, Elizabeth Fries (1857). *The Practical Housekeeper: A Cyclopedia of Domestic Economy*. New York, Stringer and Townsend.

Emery, John (1976). *European Spoons Before 1700*. Edinburgh, John Donald Publishers Ltd.

Ettlinger, Steve (1992). *The Kitchenware Book*. New York, Macmillan.

Eveleigh, David J. (1986). *Old Cooking Utensils*. Aylesbury, Shire Publications.

—— (1991). "'Put down to a Clear Bright Fire': The English Tradition of Open-Fire Roasting." *Folk Life*, 29.

Falk, Dean, and Seguchi, Noriko (2006). "Professor C. Loring Brace: Bringing Physical Anthropology ('Kicking and Screaming') into the 21st Century!" *Michigan Discussions in Anthropology*, vol. 16. pp. 175–211.

Farb, Peter, and Armelagos, George (1980). *Consuming Passions: The Anthropology of Eating*. Boston, Houghton Mifflin.

Farmer, Fannie (1896). *The Boston Cooking-School Cookbook*. Boston: Little, Brown and Company.

—— (1904). *Food and Cookery for the Sick and Convalescent*. Boston, Little, Brown and Company.

Feild, Rachel (1984). *Irons in the Fire: A History of Cooking Equipment*. Marlborough, Wiltshire, Crowood Press.

Fernández-Armesto, Felipe (2001). *Food: A History*. London, Macmillan.

Ferrie, Helke (1997). "An Interview with C. Loring Brace." *Current Anthropology*, vol. 38, no. 5, pp. 851–917.

Forbes, R. J. (1950). *Man the Maker: A History of Technology and Engineering*. London, Constable & Co.

Frederick, Christine (1916). *The New Housekeeping: Efficiency Studies in Home Management*. New York, Doubleday, Page and Company.

Friedberg, Suzanne (2009). *Fresh: A Perishable History*. Cambridge, MA, The Belknap Press of Harvard University Press.

Friedland, Susan (2009). *Vegetables: Proceedings of the Oxford Symposium on Food and Cookery 2008*. Totnes, Prospect Books.

Fuller, William (1851). *A Manual: Containing Numerous Original Recipes for Preparing Ices, With a Description of Fuller's Neapolitan Freezing Machine for Making Ices in Three Minutes at Less Expense Than Is Incurred by Any Method Now in Use*. London.

Furnivall, Frederick J., ed. (1868). *Early English Meals and Manners*. London, Kegan Paul, Trench, Truber & Co.

Gillette, Mrs. F. L., and Ziemann, Hugo (1887). *The White House Cookbook*. Washington, DC.

Gladwell, Malcolm (2010). *What the Dog Saw: And Other Adventures*. London: Penguin.

Glancey, Jonathan (2008). "Classics of Everyday Design no. 45." *Guardian*, March 25.

Goldstein, Darra (2006). Chapter on cutlery in *Feeding Desire: Design and the Tools of the Table*, edited by Sarah D. Coffin et al. New York, Assouline, in collaboration with Smithsonian Cooper-Hewitt.

Gordon, Bertram M., and Jacobs-McCusker, Lisa (1989). "One Pot Cookery and Some Comments on Its Iconography," in *Oxford Symposium on Food and Cookery 1988, The Cooking Pot: Proceedings,* pp. 55–68. Totnes, Prospect Books.

Gordon, Bob (1984). *Early Electrical Appliances.* Aylesbury, Shire Publications.

Gouffé, Jules (1874). *The Royal Book of Pastry and Confectionary,* translated from the French by Alphonse Gouffé. London, Sampson, Low, Marston.

Green, Kaye, and Leach, Foss, eds. (2007). *Vastly Ingenious: The Archaeology of Pacific Material Culture.* Dunedin, New Zealand, Otago University Press.

Green, W. C. (1922). *The Book of Good Manners: A Guide to Polite Usage.* New York, Social Mentor Publications.

Haddock, George (1954). "Frozen Foods." *Nation's Business,* June.

Hamilton, Andrew (1955). "Heat and Eat Meals Boost Frozen Food Sales." *Nation's Business,* July.

Hanawalt, Barbara (1986). *The Ties That Bound: Peasant Families in Medieval England.* New York and Oxford, Oxford University Press.

Hård, Mikael (1994). *Machines Are Frozen Spirit: The Scientification of Refrigeration and Brewing in the Nineteenth Century.* Boulder, CO, Westview Press.

Hardyment, Christina (1990). *From Mangle to Microwave: The Mechanisation of Household Work.* Cambridge, Polity.

Harland, Marion (1873). *Common Sense in the Household.* New York, Scribner, Armstrong & Co.

Harris, Gertrude (1980). *Pots and Pans.* London, Penguin.

Harrison, James, and Steel, Danielle (2006). "Burns and Scalds." *AIHW National Injury Surveillance Unit.* Flinders University, South Australia.

Harrison, Molly (1972). *The Kitchen in History.* London, Osprey.

Harrold, Charles Frederick (1930). "The Italian in Streatham Place: Giuseppe Baretti (1719–1789)." *Sewanee Review,* vol. 38, no. 2, pp. 161–175.

Harry, Karen, and Frink, Liam (2009). "The Arctic Cooking Pot: Why Was It Adopted?" *American Anthropologist,* vol. 111, pp. 330–343.

Helou, Anissa (2008). *Lebanese Cuisine.* London, Grub Street.

Herring, I. J. (1938). "The Beaker Folk." *Ulster Journal of Archaeology,* 3rd series, vol. 1, pp. 135–139.

Hertzmann, Peter (2007). *Knife Skills Illustrated: A User's Manual.* New York, W. W. Norton.

Hess, Karen, ed. (1984). *The Virginia House-wife by Mary Randolph.* Columbia, University of South Carolina Press.

Hesser, Amanda (2005). "Under Pressure." *New York Times,* August 14.

Hessler, Martina (2009). "The Frankfurt Kitchen: The Model of Modernity and the 'Madness' of Traditional Users, 1926 to 1933," in *Cold War Kitchen,* edited by Ruth Oldenziel and Karin Zachmann. Cambridge, MIT Press.

Homer, Ronald F. (1975). *Five Centuries of Base Metal Spoons*. London, The Worshipful Company of Pewterers.

Homes, Rachel (1973). "Mixed Blessings of a Food Mixer." *London Times*, August 9.

Hoopes, John W., and Barnett, William K., eds. (1995). *The Emergence of Pottery: Technology and Innovation in Ancient Societies*. Washington, DC, Smithsonian Institution Press.

Hosking, Richard (1996). *A Dictionary of Japanese Food*. Totnes, Prospect Books.

Hughes, Bernard, and Hughes, Therle (1952). *Three Centuries of English Domestic Silver, 1500–1820*. London, Lutterworth Press.

Hutchinson, R. C. (1966). *Food Storage in the Home*. London, Edward Arnold.

Isenstadt, Sandy (1998). "Visions of Plenty: Refrigerators in America Around 1950." *Journal of Design History*, vol. 11, no. 4, pp. 311–321.

Ishige, Naomichi (2001). *The History and Culture of Japanese Food*. London, Kegan Paul.

Jaine, Tom, ed. (1989). *Oxford Symposium on Food and Cookery 1988, The Cooking Pot: Proceedings*. Totnes, Prospect Books.

Jay, Sarah (2008). *Knives Cooks Love: Selection, Care, Techniques, Recipes*. Kansas City, Andrew McMeel.

Kafka, Barbara (1987). *Microwave Gourmet*. New York, William Morrow.

Kalm, Pehr (1892). *Kalm's Account of His Visit to England on His Way to America in 1748*, translated by Joseph Lucas. London, Macmillan.

Keller, Thomas (2008). *Under Pressure: Cooking Sous Vide*. New York, Artisan Books.

Kinchin, Juliet, with O'Connor, Aidan (2011). *Counter Space: Design and the Modern Kitchen*. New York, Museum of Modern Art.

Kitchiner, William (1829). *The Cook's Oracle and Housekeeper's Manual*, 3rd ed. Edinburgh, A. Constable & Co.

Koerner, Brendan I. (2006). "A Spork with Added Edge." *New York Times*, September 17.

Koon, H. E. C., O'Connor, T. P., and Collins, M. J. (2010). "Sorting the Butchered from the Boiled." *Journal of Archaeological Science*, vol. 37, pp. 62–69.

Kranzberg, Melvin (1986). "Technology and History: Kranzberg's Laws." *Technology and Culture*, vol. 27, June, pp. 544–560.

Kurti, Nicholas, and Kurti, Giana, eds. (1988). *But the Crackling Is Superb: An Anthology on Food and Drink by Fellows and Foreign Members of the Royal Society*. Bristol, Hilger.

Lamb, Charles (2011). *A Dissertation upon Roast Pig and Other Essays*. London, Penguin.

Larner, John W. (1986). "Judging the Kitchen Debate." *OAH Magazine of History*, vol. 2, no. 1, pp. 25–26.

Larson, Egon (1961). *A History of Invention*. London, Phoenix House.

Lawrence, Keith (2010). "Costs Add Up for Jail's 'Sporks,' Other Items." *McClatchy-Tribune Business News*, July 6.

Leach, Helen M. (1982). "Cooking Without Pots: Aspects of Prehistoric and Traditional Polynesian Cooking." *New Zealand Journal of Archaeology*, vol. 4, pp. 149–156.

———— (2007). "Cooking with Pots—Again," in *Vastly Ingenious: The Archaeology of Pacific Material Culture*, edited by Atholl Anderson, Kaye Green, and Foss Leach, pp. 53–68. Dunedin, New Zealand, Otago University Press.

Lemme, Chuck (1989). "The Ideal Pot," in *Oxford Symposium on Food and Cookery 1988, The Cooking Pot: Proceedings*, pp. 82–99. Totnes, Prospect Books.

Lincoln, Mary Johnson (1884). *Mrs. Lincoln's Boston Cook Book: What to Do and What Not to Do in Cooking*. Boston, Roberts Brothers.

Lloyd, G. I. K. (1913). *The Cutlery Trades: An Historical Essay in the Economics of Small-Scale Production*. London, Longmans Green & Co.

Lockley, Lawrence C. (1938). "The Turn-Over of the Refrigerator Market." *Journal of Marketing*, vol. 2, no. 3, pp. 209–213.

MacDonald, John (1985). *Memoirs of an Eighteenth-Century Footman*. London, Century.

MacGregor, Neil (2010). *A History of the World in 100 Objects*. London, Allen Lane.

Mackenzie, Donald, and Wajcman, Judy, eds. (1985). *The Social Shaping of Technology: How the Refrigerator Got Its Hum*. Milton Keynes, Open University Press.

Man, Edward Horace (1932). *On the Aboriginal Inhabitants of the Andaman Islands*. London, Royal Anthropological Institute.

Marquardt, Klaus (1997). *Eight Centuries of European Knives, Forks and Spoons*, translated by Joan Clough. Stuttgart, Arnoldsche.

Marsh, Stefanie (2003). "Can't Cook. Won't Cook. Don't Care. Going Out." *London Times*, November 17.

Marshall, Mrs. A. B. (1857). *The Book of Ices*, 2nd ed. London, Marshall's School of Cookery.

———— (1894). *Fancy Ices*. London, Simpkin, Hamilton & Kent & Co.

———— (1896). *Mrs. A.B. Marshall's Cookery Book*. London, Simpkin, Hamilton & Kent & Co.

Marshall, Jo (1976). *Kitchenware*. London, BPC Publishers.

Martino, Maestro (2005). *The Art of Cooking, Composed by the Eminent Maestro Martino of Como*, edited by Luigi Ballerini, translated by Jeremy Parzen. Berkeley, London, University of California Press.

Masters, Thomas (1844). *The Ice Book*. London, Simpkin, Marshall & Co.

May, Robert (2000). *The Accomplisht Cook or the Art and Mystery of Cookery, a Facsimile of the 1685 Edition*, edited by Alan Davidson, Marcus Bell, and Tom Jaine. Totnes, Prospect Books.

McEvedy, Allegra (2011). *Bought, Borrowed and Stolen: Recipes and Knives from a Travelling Chef*. London, Conran Octopus.

McGee, Harold (1986). *On Food and Cooking: The Science and Lore of the Kitchen*. London, Allen and Unwin.

McNeil, Ian, ed. (1990). *An Encyclopedia of the History of Technology*. London, Routledge.

Mellor, Maureen (1997). *Pots and People That Have Shaped the Heritage of Medieval and Later England*. Oxford, Ashmolean Museum.

Mintel (1998). "Microwave Ovens." Mintel Report, UK.

Myers, Lucas (1989). "Ah, Youth . . . : Ted Hughes and Sylvia Plath at Cambridge and After." *Grand Street*, vol. 8, no. 4.

Myhrvold, Nathan (2011). *Modernist Cuisine: The Art and Science of Cooking*, 6 vols. Bellevue, WA, The Cooking Lab.

Nakano, Yoshiko (2010). *Where There Are Asians, There Are Rice Cookers*. Hong Kong, Hong Kong University Press.

Nasrallah, Nawal, ed. (2007). *Annals of the Caliph's Kitchen: Translation with Introduction and Glossary*. Leiden, Brill.

Newman, Barry (2009). "To Keep the Finger out of Finger Food, Inventors Seek a Better Bagel Cutter." *Wall Street Journal*, December 1.

Nickles, Shelley (2002). "Preserving Women: Refrigerator Design as Social Process in the 1930s." *Technology and Culture*, vol. 43, no. 4, pp. 693–727.

O'Connor, Desmond (2004). "Baretti, Giuseppe Marc'Antonio (1719–1789)." *Oxford Dictionary of National Biography*. Oxford, Oxford University Press.

Ohren, Magnus (1871). *On the Advantages of Gas for Cooking and Heating*. London, printed for The Crystal Palace District Gas Company.

Oka, K., Sakuarae, A., Fujise, T., Yoshimatzu, H., Sakata, T., and Nakata, M. (2003). "Food Texture Differences Affect Energy Metabolism in Rats." *Journal of Dental Research*, vol. 82, June 2003, pp. 491–494.

Oldenziel, Ruth, and Zachmann, Karin, eds. (2009). *Cold War Kitchen: Americanization, Technology, and European Users*. Cambridge, MA, MIT Press.

Ordway, Edith B. (1918). *The Etiquette of Today*. New York, Sully & Kleinteich.

Osepchuk, John M. (1984). "A History of Microwave Heating." *IEEE Transactions on Microwave Theory and Techniques*, vol. 32, no. 9, pp. 1200–1224.

——— (2010). "The Magnetron and the Microwave Oven: A Unique and Lasting Relationship." *Origins and Evolution of the Cavity Magnetron (CAVMAG)*, *2010 International Conference*, April 19–20, pp. 46–51.

Owen, Sri (2008). *Sri Owen's Indonesian Food*. London, Pavilion.

Parloa, Maria (1882). *Miss Parloa's New Cookbook*. New York, C. T. Dillingham.

Parr, Joy (2002). "Modern Kitchen, Good Home, Strong Nation." *Technology and Culture*, vol. 43, no. 4, pp. 657–667.

Pierce, Christopher (2005). "Reverse Engineering the Ceramic Cooking Pot: Cost and Performance Properties of Plain and Textured Vessels." *Journal of Archaeological Method and Theory*, vol. 12, no. 2, pp. 117–157.

Plante, Ellen M. (1995). *The American Kitchen: From Hearth to Highrise*. New York, Facts on File.

Pollan, Michael (2008). *In Defense of Food: An Eater's Manifesto*. New York, Penguin Press.

———— (2009). "Out of the Kitchen, onto the Couch." *New York Times*, August 2.

Post, Emily (1960). *The New Emily Post's Etiquette*. New York, Funk and Wagnalls.

Potter, Jeff (2010). *Cooking for Geeks: Real Science, Great Hacks, and Good Food*. Sebastopol, CA, O'Reilly Media.

Power, Eileen, ed. (1992). *The Goodman of Paris (Le Ménagier de Paris, c. 1393)*, translated by Eileen Power. London, Folio Society.

Pufendorf, Samuel (1695). *An Introduction to the History of the Principal Kingdoms and States of Europe*. London, M. Gilliflower.

Quennell, Marjorie, and Quennell, C. H. B. (1957 [first published 1918]). *A History of Everyday Things in England, Volume 1, 1066–1499*. London, B. T. Batsford.

Randolph, Mary (1838). *The Virginia Housewife or Methodical Cook*. Baltimore, Plaskitt, Fite.

Rath, Eric C. (2010). *Food and Fantasy in Early Modern Japan*. Berkeley, University of California Press.

Reid, Susan (2002). "Cold War in the Kitchen: Gender and the De-Stalinization of Consumer Taste in the Soviet Union Under Khrushchev." *Slavic Review*, vol. 61, no. 2, pp. 211–252.

———— (2005). "The Khrushchev Kitchen: Domesticating the Scientific-Technological Revolution." *Journal of Contemporary History*, vol. 40, no. 2, pp. 289–316.

———— (2009). "'Our Kitchen Is Just as Good': Soviet Responses to the American Kitchen," in *Cold War Kitchen: Americanization, Technology, and European Users*, edited by Ruth Oldenziel and Karin Zachmann. Cambridge, MA, MIT Press.

Renton, Alex (2010). "Sous-Vide Cooking: A Kitchen Revolution." *Guardian*, September 2.

Rios, Alicia (1989). "The Pestle and Mortar," in *Oxford Symposium on Food and Cookery 1988, The Cooking Pot: Proceedings*, pp. 125–136. Totnes, Prospect Books.

Rodgers, Judy (2002). *The Zuni Café Cookbook*. New York, W. W. Norton.

Rogers, Ben (2003). *Beef and Liberty: Roast Beef, John Bull and the English Nation*. London, Chatto & Windus.

Rogers, Eric (1997). *Making Traditional English Wooden Eating Spoons*. Felixstowe, Woodland Craft Supplies.

Rorer, Sarah Tyson (1902). *Mrs Rorer's New Cookbook*. Philadelphia, Arnold & Co.

Ross, Alice (2007). "Measurements," in *The Oxford Companion to American Food and Drink*, edited by Andrew F. Smith. Oxford, Oxford University Press.

Routledge, George (1875). *Routledge's Manual of Etiquette*. London, New York, George Routledge & Sons.

Ruhlman, Michael (2009). *Ratio: The Simple Codes Behind the Craft of Everyday Cookery*. New York, Scribner Book Company.

Rumford, Benjamin, Count von (1968). *Collected Works of Count Rumford*, edited by Sanborn Brown. Cambridge, MA, Harvard University Press.

Salisbury, Harrison E. (1959). "Nixon and Khrushchev Argue in Public as U.S. Exhibit Opens." *New York Times*, July 25.

Samuel, Delwen (1999). "Bread Making and Social Interactions at the Amarna Workmen's Village, Egypt." *World Archaeology*, vol. 31, no. 1, pp. 121–144.

Sanders, J. H. (2000). "Nicholas Kurti C.B.E." *Biographical Memoirs of Fellows of the Royal Society*, vol. 46, pp. 300–315.

Scappi, Bartolomeo (2008). *The Opera of Bartolomeo Scappi (1570)*, translated with commentary by Terence Scully. Toronto, University of Toronto Press.

Scully, Terence (1995). *The Art of Cookery in the Late Middle Ages*. Woodbridge, Boydell.

Segre, Gino (2002). *Einstein's Refrigerator: Tales of the Hot and Cold*. London, Allen Lane.

Seneca, Lucius Annaeus (2007). *Dialogues and Essays*, translated by John Davie. Oxford, Oxford University Press.

Serventi, Silvano, and Sabban, Françoise (2002). *Pasta: The Story of a Universal Food*, translated by Anthony Shugaar. New York, Columbia University Press.

Shapiro, Laura (1986). *Perfection Salad: Women and Cooking at the Turn of the Century*. New York, Farrar, Straus and Giroux.

Shephard, Sue (2000). *Pickled, Potted and Canned: The Story of Food Preserving*. London, Headline.

Shleifer, Andrei, and Treisman, Daniel (2005). "A Normal Country: Russia After Communism." *Journal of Economic Perspectives*, vol. 19, no. 1, pp. 151–174.

Simmons, Amelia (1798). *American Cookery*. Hartford, Simeon Butler.

Smith, Andrew F., ed. (2007). *The Oxford Companion to American Food and Drink*, 2 vols. Oxford, Oxford University Press.

——— (2009). *Eating History: Thirty Turning Points in the Making of American Cuisine*. New York, Columbia University Press.

Snodin, Michael (1974). *English Silver Spoons*. London, Charles Letts & Company.

So, Yan-Kit (1992). *Classic Food of China*. London, Macmillan.

Sokolov, Ray (1989). "Measure for Measure," in *Oxford Symposium on Food and Cookery 1988, The Cooking Pot: Proceedings*, pp. 148–152. Totnes, Prospect Books.

Soyer, Alexis (1853). *The Pantropheon or History of Food and Its Preparation from the Earliest Ages of the World*. London, Simpkin, Marshall & Co.

Sparkes, B. A. (1962). "The Greek Kitchen." *Journal of Hellenic Studies*, vol. 82, pp. 121–137.

Spencer, Colin (2002). *British Food: An Extraordinary Thousand Years of History.* London, Grub Street.

———— (2011). *From Microliths to Microwaves.* London, Grub Street.

————, and Clifton, Claire (1993). *The Faber Book of Food.* London, Faber and Faber.

Spurling, Hilary, ed. (1986). *Elinor Fettiplace's Receipt Book.* London, Viking Salamander.

Standage, Tom (2009). *An Edible History of Humanity.* London, Atlantic Books.

Stanley, Autumn (1993). *Mothers and Daughters of Invention: Notes for a Revised History of Technology.* London, Scarecrow Press.

Strong, Roy (2002). *Feast: A History of Grand Eating.* London, Jonathan Cape.

Sugg, Marie Jenny (1890). *The Art of Cooking by Gas.* London, Cassell.

Sydenham, P. H. (1979). *Measuring Instruments: Tools of Knowledge and Control.* London, Peter Peregrinus.

Symons, Michael (2001). *A History of Cooks and Cooking.* Totnes, Prospect Books.

Tannahill, Reay (2002). *Food in History* (new and updated edition). London, Review.

Tavernor, Robert (2007). *Smoot's Ear: The Measure of Humanity.* New Haven, Yale University Press.

Teaford, Mark, and Ungar, Peter (2000). "Diet and the Evolution of the Earliest Human Ancestors." *Proceedings of the National Academy of Sciences of the United States of America*, vol. 97, no. 25, pp. 13,506–13,511.

This, Hervé (2005). "Molecular Gastronomy." *Nature Materials*, vol. 4, pp. 5–7.

———— (2009). *The Science of the Oven.* New York, Columbia University Press.

Thoms, Alston V. (2009). "Rocks of Ages: Propagation of Hot-Rock Cookery in Western North America." *Journal of Archaeological Science*, vol. 36, pp. 573–591.

Thornton, Don (1994). *Beat This: The Eggbeater Chronicles.* Sunnyvale, Offbeat Books.

Toomre, Joyce, ed. (1992). *Classic Russian Cooking: Elena Molokhovets' A Gift to Young Housewives.* Bloomington, Indiana University Press.

Toth, Nicholas, and Schick, Kathy (2009). "The Oldowan: The Tool Making of Early Hominins and Chimpanzees Compared." *Annual Review of Anthropology*, vol. 38, pp. 289–305.

Toussaint-Samat, Maguelonne (1992). *A History of Food*, translated by Anthea Bell. Oxford, Blackwell Reference.

Trager, James (1996). *The Food Chronology.* London, Aurum Press.

Trevelyan, G. M. (1978 [first published 1944]). *English Social History: A Survey of Six Centuries from Chaucer to Queen Victoria.* London, Longman.

Troubridge, Lady (1926). *The Book of Etiquette*, 2 vols. London, The Associated Bookbuyer's Company.

Unger, Richard W. (1980). "Dutch Herring, Technology and International Trade in the Seventeenth Century." *Journal of Economic History*, vol. 40, no. 2, pp. 253–280.

Visser, Margaret (1991). *The Rituals of Dinner: The Origins, Evolution, Eccentricities and Meaning of Table Manners*. London, Penguin Books.

Vitelli, Karen D. (1989). "Were Pots First Made for Food? Doubts from Franchti." *World Archaeology*, vol. 21, no. 1, pp. 17–29.

—— (1999). "'Looking Up' at Early Ceramics in Greece," in *Pottery and People: A Dynamic Interaction*, edited by James M. Skibo and Gary M. Feinman, pp. 184–198. Salt Lake City, University of Utah Press.

Waines, David (1987). "Cereals, Bread and Society: An Essay on the Staff of Life in Medieval Iraq." *Journal of the Economic and Social History of the Orient*, vol. 30, no. 3, pp. 255–285.

Wandsnider, LuAnn (1997). "The Roasted and the Boiled: Food Consumption and Heat Treatment with Special Emphasis on Pit-Hearth Cooking." *Journal of Anthropological Archaeology*, vol. 16, pp. 1–48.

Weber, Robert J. (1992). *Forks, Phonographs and Hot Air Balloons: A Field Guide to Inventive Thinking*. Oxford, Oxford University Press.

Webster, Thomas (1844). *An Encyclopedia of Domestic Economy*. London, Longman, Brown, Green and Longmans.

Weinstein, Rosemary (1989). "Kitchen Chattels: The Evolution of Familiar Objects 1200–1700," in *Oxford Symposium on Food and Cookery 1988, The Cooking Pot: Proceedings*, pp. 168–183. Totnes, Prospect Books.

Weir, Robin, and Weir, Caroline (2010). *Ices, Sorbets and Gelati: The Definitive Guide*. London, Grub Street.

Weir, Robin, Brears, Peter, Deith, John, and Barham, Peter (1998). *Mrs. Marshall: The Greatest Victorian Ice Cream Maker with a Facsimile of the Book of Ices*. Smith Settle, Syon House.

Wheaton, Barbara (1983). *Savouring the Past: The French Kitchen and Table from 1300 to 1789*. London, Chatto & Windus.

Whitelaw, Ian (1997). *A Measure of All Things: The Story of Measurement Through the Ages*. Hove, England, David & Charles.

Wilkins, J. (1680). *Mathematical Magick or the Wonders that May Be Performed by Mechanical Geometry*. London, Edward Gellibrand.

Wilkinson, A. W. (1944). "Burns and Scalds in Children: An Investigation of their Cause and First-Aid Treatment." *British Medical Journal*, vol. 1, no. 4331, pp. 37–40.

Wilson, C. Anne (1973). *Food and Drink in Britain from the Stone Age to Recent Times*. London: Constable.

Wolf, Burt (2000). *The New Cooks' Catalogue*. New York, Alfred A. Knopf.

Wolfman, Peri, and Gold, Charles (1994). *Forks, Knives and Spoons.* London, Thames & Hudson.

Woodcock, F. Huntley, and Lewis, W. R. (1938). *Canned Foods and the Canning Industry.* London, Sir I. Pitman.

Woolley, Hannah (1672). *The Queen-Like Closet or Rich Cabinet.* London, Rich. Lownes.

———— (1675). *The Accomplish'd Lady's Delight in Preserving, Physick, Beautifying, and Cookery.* N.p.

Worde, Wynkyn de (2003). *The Boke of Keruynge* (The Book of Carving), with an introduction by Peter Brears. Lewes, Southover Press.

Wrangham, Richard (2009). *Catching Fire: How Cooking Made Us Human.* London, Profile.

————, with Jones, James Holland, Laden, Greg, Pilbeam, David, and Conklin-Brittain, Nancy Lou (1999). "The Raw and the Stolen." *Current Anthropology,* vol. 40, no. 5, pp. 567–594.

Wright, Katherine (1994). "Ground-Stone Tools and Hunter-Gatherer Subsistence in Southwest Asia: Implications for the Transition to Farming." *American Antiquity,* vol. 59, no. 2, pp. 238–263.

Yarwood, Doreen (1981). *British Kitchen: Housewifery Since Roman Times.* London, Batsford.

Young, Carolin (2002). *Apples of Gold in Settings of Silver: Stories of Dinner as a Work of Art.* London, Simon and Schuster.

———— (2006). "The Sexual Politics of Cutlery," in *Feeding Desire: Design and the Tools of the Table,* edited by Sarah D. Coffin et al. New York, Assouline, in collaboration with Smithsonian Cooper-Hewitt,

Young, H. M. (1897). *Domestic Cooking with Special Reference to Cooking by Gas,* 21st edition. Chester, H. M. Young.

INDEX

A BASIC BOOKS READING GROUP GUIDE TO

CONSIDER THE FORK

Questions for Discussion

1. Bee Wilson covers a wide range of culinary tools in *Consider the Fork*, but obviously she could not include everything. Are there any tools that you wish had been included in this book, but weren't? More generally, what tools continue to be underappreciated or underused—and do these share any characteristics with the ones that Wilson *does* discuss in the book?

2. Put together a list of must-have kitchen tools. Where do you draw the line between essential utilities that no cook can operate without, and frivolous accessories that just clutter up the kitchen?

3. In the chapter "Grind," Wilson writes that the Cuisinart "transform[ed] cooking from pain to pleasure"—and yet she confesses to still using "obsolete" or needlessly labor-intensive technologies, such as when she grinds basil and garlic with a mortar and pestle to make pesto. Do you ever choose to make food in an old-fashioned way, even when more advanced tools are available? If so, why?

4. Is your appreciation for food diminished when it's prepared quickly and relatively effortlessly? Would you enjoy butter more if you knew that hours of human labor went into it, or would meringues be more delectable if you knew that multiple people had tired themselves out to stiffen the egg whites? Is speed and convenience a luxury when it comes to cooking and eating, or is "slow food" more of a pleasure?

5. Can you think of utensils that were once widely used, but which have since disappeared from most kitchens? Did any of these items work better than the ones that replaced them, in your opinion?

6. Even after the development of indoor gas ranges, cooks clung to the time-tried method of hearth cooking despite its many dangers. Are there any technologies that people continue to use today, despite the obvious risks associated with them?

7. One of Wilson's arguments is that culinary technologies have shaped the human body, in ways that we often don't think of; the alignment of our teeth, for instance, is intimately connected to our use of knives. Are there other kitchen tools that have had an impact on our bodies? On our minds?

8. The tools that we use to present and consume food can be every bit as central to our experience of eating as the food itself: few people would choose to eat sushi with a fork, say, or to drink expensive wine out of a paper cup. Think of your favorite food: does it have a specific utensil associated with it? If so, what does that utensil say about you—either the personal culinary traditions you have inherited, or those that you have chosen for yourself?

9. Wilson's book ends with a look at molecular gastronomy, and the scientific tools and techniques of "modernist" chefs around the world. Is modernist cooking a passing trend, or are some of the tools it has generated—the sous-vide machine, for instance—here to stay?

10. "The 'kitchen of tomorrow,'" Wilson writes, "was a staple of twentieth-century life." What does the "kitchen of tomorrow" look like, at the dawn of the third millennium? Are there certain tools, techniques, or kitchen design elements that will never go out of style, no matter how many years go by?